Aviation Mechanic

Oral & Practical Exam Guide

Comprehensive preparation for the FAA Aviation Mechanic General, Airframe, and Powerplant Oral & Practical Exams

Fifth Edition

Keith Anderson

Based on original text by Dale Crane

AVIATION SUPPLIES & ACADEMICS, INC.
KALAMAZOO, MICHIGAN

Aviation Mechanic Oral & Practical Exam Guide
Fifth Edition
by Keith Anderson
Based on original text by Dale Crane

Aviation Supplies & Academics, Inc.
817 Walbridge Street
Kalamazoo, Michigan 49007
asa@asa2fly.com | 425-235-1500 | asa2fly.com

First edition published 1994. Fifth edition published 2024.

See **asa2fly.com/oegamt** for additional information and updates related to this book.

ASA-OEG-AMT5
ISBN 978-1-64425-362-5

Additional formats available:
eBook EPUB ISBN 978-1-64425-363-2
eBook PDF ISBN 978-1-64425-364-9

Printed in the United States of America
2029 2028 2027 2026 9 8 7 6

Cover photo: Juice Flair/Shutterstock.com

Library of Congress Cataloging-in-Publication Data:
Names: Anderson, Keith, 1960- author. | Crane, Dale, author.
Title: Aviation mechanic oral & practical exam guide : comprehensive preparation for the FAA aviation mechanic general, airframe, and powerplant oral & practical exams / Keith Anderson, original text by Dale Crane.
Other titles: Oral & practical exam guide : comprehensive preparation for the FAA aviation mechanic general, airframe, and powerplant oral & practical exams
Description: Fifth edition. | Newcastle, Washington : Aviation Supplies & Academics, Inc., 2024. | "ASA-OEG-AMT5".—Title page verso.
Identifiers: LCCN 2023041443 (print) | LCCN 2023041444 (ebook) | ISBN 9781644253625 (trade paperback) | ISBN 9781644253632 (epub) | ISBN 9781644253649 (pdf)
Subjects: LCSH: United States. Federal Aviation Administration—Examinations—Study guides. | Airplanes—Maintenance and repair—Examinations, questions, etc. | Aviation mechanics (Persons)—Certification—United States. | Airplanes—Motors—Maintenance and repair—Examinations, questions, etc. | LCGFT: Examinations.
Classification: LCC TL671.9 .C66468 2024 (print) | LCC TL671.9 (ebook) | DDC 629.134/6—dc23/eng/20231002
LC record available at https://lccn.loc.gov/2023041443
LC ebook record available at https://lccn.loc.gov/2023041444

[03]

Contents

The Powerplant Oral and Practical Tests 173

Preface

Certification as an aviation mechanic is a major step in your career. You were required to have a certain level of experience to qualify to take your knowledge tests, and having passed all sections, you are now ready for the final step, the oral and practical tests.

The knowledge tests are strictly objective and verify only your knowledge of facts. When taking them, you are interfacing with a computer and there is no personal involvement.

The oral and practical tests are different. In these you work directly with an experienced mechanic on a one-on-one basis. This examiner is able to not only judge your mechanical skills, but to observe the way you think and see the way you solve problems.

It is important that you approach the oral and practical tests with the proper mental attitude. The examiner has one basic thought in mind, that of determining whether or not you have the level of knowledge, skill, and risk management needed for an entry-level aviation mechanic. The examiner will not try to trick you in any way, and he or she wants you to pass almost as much as you do.

If you are asked a question to which you do not know the answer, admit it rather than try to bluff your way through. If you are given a project that you do not understand, discuss it with the examiner. In particular, avoid stumbling through a project that you lack the ability to execute properly. The examiner will discuss the project with you but will show little or no tolerance for your driving ahead with a project you obviously cannot handle.

The oral and practical tests are your last steps toward certification, so study this guide carefully as it is designed to help you prepare for them. We wish you success.

Dale Crane, 2000

Preface to the Fifth Edition

The Oral and Practical Exam process is now guided by the Airman Certification Standards (ACS). The ACS contains many of the same knowledge subjects and skills projects as the previous Practical Test Standards (PTS), but there is less emphasis on the demonstration of skills that have become specialized trades such as welding, wood structures, and fabric covering. The knowledge section of the exam now includes risk management questions for which the student is expected to understand the risks associated with certain tasks and explain the proper ways to mitigate those risks.

All oral and practical questions and projects are randomly generated from the FAA test generator, and the list of questions and projects are then given to the examiner. Since the examiner does not know which questions and projects will be assigned, both the examiner and the student must be prepared for all possible projects. Applicants should expect to be retested on subjects they missed during the knowledge exam; the subjects associated with the codes on your Airman Knowledge Test Report will be retested during the Oral and Practical Exam.

Keith Anderson

Certification of Maintenance Airmen

The Federal Aviation Administration has three classifications of maintenance airmen: repairman, authorized inspector, and aviation mechanic. Certification in each category has special requirements and special privileges. This Oral & Practical Exam Guide applies to the tests for Aviation Mechanic certification, but all three classifications are described below.

Repairman

The applicant for a Repairman Certificate must be employed for a specific job requiring his or her special qualifications by a certificated commercial operator or certificated air carrier.

A repairman applicant must have at least 18 months of practical experience in the procedures, practices, inspection methods, materials, tools, machine tools, and equipment generally used in the maintenance duties of the specific job for which he or she is to be employed and certificated. Alternatively, the applicant may complete specialized formal training that is acceptable to the administrator and specifically designed to qualify the applicant for the job for which he or she is to be employed.

A repairman may exercise the privileges of the certificate only in connection with the duties for the certificate holder by whom the repairman was certificated and recommended. There is a special type of repairman certificate issued to the builder of an experimental aircraft which allows the holder to perform condition inspections on the aircraft constructed by him or her.

Authorized Inspector

An applicant for an Inspection Authorization (IA) must:

- Hold a currently effective Aviation Mechanic Certificate with both Airframe and Powerplant Ratings that has been in effect for a total of at least 3 years.
- Have been actively engaged, for at least the 2-year period before the date of application, in maintaining civil certificated aircraft.
- Have a fixed base of operation.
- Have available the equipment, facilities, and inspection data necessary to properly inspect airframes, powerplants, propellers, or any related part or appliance.
- Pass a knowledge test on his or her ability to inspect according to safety standards for returning aircraft to service after major repairs and major alterations, and annual and progressive inspections performed under 14 CFR Part 43.

The holder of an Inspection Authorization may:

- Inspect and approve for return to service an aircraft after a major repair or major alteration if the work has been done in accordance with technical data that has been approved by the administrator.
- Perform an annual inspection, or perform or supervise a progressive inspection.

An Inspection Authorization expires on March 31 of each odd-numbered year and must be renewed for a two-year period at that time.

Aviation Mechanic

The FAA issues an Aviation Mechanic Certificate with an Airframe Rating, Powerplant Rating, or both ratings to applicants who are properly qualified. Below are descriptions of the experience, knowledge, and practical requirements and suggested study references for all three ratings.

Requirements for Mechanic Certification

14 CFR Part 65, Certification: Airmen Other Than Flight Crewmembers, covers the requirements for mechanic certification, described below.

Basic Requirements

- Must be at least 18 years of age.
- Must be able to read, write, speak, and understand the English language, or in the case of an applicant who does not meet this requirement and who is employed outside of the United States by a U.S. air carrier, have his or her certificate endorsed "Valid only outside the United States."
- Must have passed all of the prescribed tests within a period of 24 months.

Experience Requirements

Must have a graduation certificate or certificate of completion from a certificated aviation maintenance technician school, or documentary evidence, satisfactory to the Administrator, of:

- At least 18 months of practical experience with the procedures, practices, materials, tools, machine tools, and equipment generally used in constructing, maintaining, or altering airframes or powerplants appropriate to the rating sought; or
- At least 30 months of practical experience concurrently performing the duties appropriate to both the airframe and powerplant ratings.

Knowledge Requirements and Knowledge Tests

After meeting the applicable experience requirements, each applicant for an Aviation Mechanic Certificate must pass a knowledge test covering the construction and maintenance of aircraft appropriate to the rating sought, the regulations that pertain to the rating, and the applicable provisions of 14 CFR Part 43 (Maintenance, Preventive Maintenance, Rebuilding, and Alteration) and Part 91 (General Operating and Flight Rules).

The basic principles covering the installation and maintenance of propellers are included in the Powerplant test.

The applicant must pass each section of the knowledge test before applying for the oral and practical tests. There are three knowledge tests, a General test that is required for both ratings, and tests for both the Airframe and Powerplant Ratings. An applicant for the Airframe Rating must pass the General and the Airframe test, and an applicant for the Powerplant Rating must pass the General and Powerplant test. The General test needs to be taken only one time. All test questions are the objective, multiple-choice type with three choices of answers. The minimum passing grade for each test is 70 percent.

The General test consists of 60 multiple-choice questions selected by computer from more than 600 questions in the Aviation Mechanic—General test question bank. You are allowed 2 hours to take this test. The Airframe and Powerplant tests each consist of 100 multiple-choice questions taken from the more than 1,000-question Aviation Mechanic—Airframe and the more than 1,000-question Aviation Mechanic—Powerplant test question banks. You are allowed 2 hours for each of these tests.

If the score on your Airman Knowledge Test Report is 70 or above, the report is valid for 24 calendar months. You may elect to retake the test in anticipation of a better score, after 30 days from the date your test was taken. The score of the latest test you take will become the official test score. If you fail a knowledge test, you may apply for retesting before 30 days if you present the failed test report and an endorsement from an authorized Aviation Mechanic Certificate holder. This endorsement must certify that additional instruction has been given and that you have been found competent to pass the test (the endorsement is not necessary if you wait 30 days).

Skill Requirements

Each applicant for an Aviation Mechanic Certificate or Rating must pass an oral and a practical test on the rating sought by demonstrating the assigned objectives for the relevant subject areas contained in the *Aviation Mechanic General, Airframe, and Powerplant Airman Certification Standards* (FAA-S-ACS-1). The oral and practical portion of the tests assesses the applicant's application of the knowledge, risk management, and skill in these subject areas. These testing procedures are covered in detail beginning on page x.

An applicant for a Powerplant Rating must show his or her ability to make satisfactory minor repairs to and minor alterations of propellers.

The examiner will download an oral and practical examination that is generated at random for each applicant. Each candidate should be familiar with all the knowledge, risk management, and skill requirements contained within the appropriate airman certification standards.

How to Use This Guide

The ASA Test Guides for General, Airframe, and Powerplant Mechanic have been specially prepared to help you get ready to take your FAA knowledge tests. The same material is covered in your oral and practical tests, so it is important to review all of the questions and answers in the knowledge test portion of these Guides in preparation for your O&P.

The questions in this *Aviation Mechanic Oral & Practical Exam Guide* are typical of those you will likely be asked. The Skills section in each subject area includes the practical projects that are typical of those the designated mechanic examiner (DME) will be apt to use to check your level of skill. The *actual* questions and projects will be chosen at random by the FAA test generator, to include retesting knowledge proven deficient on the Knowledge Exams, and will be given to your DME to be used during the evaluation.

Your examiner is a knowledgeable mechanic who can evaluate your capabilities, so don't try to "snow" the examiner with words when you don't know the answer, and don't attempt any project that you are not competent to handle. It is far better to admit your lack of knowledge or skill than to blunder into a project and show that you lack the judgment to properly evaluate your capabilities.

All questions and skill elements in this guide are followed by the corresponding ACS code (e.g., AM.I.A.K1) that they align with to aid you in preparing for the tests. See a more detailed explanation of ACS element codes under the "Test Standards" section on page x.

Reference Codes Used in this Guide

Included in each section of this guide are references to other ASA resources, FAA handbooks, FAA Advisory Circulars (ACs), Federal Aviation Regulations, and other study materials that apply specifically to that section or subject covered. These resources as well as additional documents that are valuable in preparing you for your oral and practical exam are listed below.

14 CFR	*Title 14 of the Code of Federal Regulations* (Applicable parts are available in ASA's FAR-AMT)
AC 25-11	*Electronic Flight Displays*
AC 25.1455-1	*Waste Water/Potable Water Drain System Certification Testing*
AC 39-7	*Airworthiness Directives*
AC 43-4	*Corrosion Control for Aircraft*
AC 43-9	*Maintenance Records*
AC 43.13-1	*Acceptable Methods, Techniques, and Practices—Aircraft Inspection and Repair*
AC 43.13-2	*Acceptable Methods, Techniques, and Practices—Aircraft Alterations*
AC 43-215	*Standardized Procedures for Performing Aircraft Magnetic Compass Calibration*
AC 45-2	*Identification and Registration Marking*
AC 60-11	*Test Aids and Materials that May be Used by Airman Knowledge Testing Applicants*
AC 60-28	*FAA English Language Standard for an FAA Certificate Issued Under 14 CFR Parts 61, 63, 65, and 107*
AC 65-2	*Airframe and Powerplant Mechanic's Certification Guide*
AC 120-39	*Hazards of Waste Water Ice Accumulation Separating from Aircraft in Flight*
AC 120-72	*Maintenance Human Factors Training*
AC 150/5210-20	*Ground Vehicle Operations to include Taxiing or Towing an Aircraft on Airports*
ASA-MHB	*Aviation Mechanic Handbook* (ASA)
DOT/FAA/AM-11/10	*Fatigue Risk Management in Aviation Maintenance: Current Best Practices and Potential Future Countermeasures*
FAA-H-8083-1	*Aircraft Weight and Balance Handbook*
FAA-H-8083-2	*Risk Management Handbook*
FAA-H-8083-25	*Pilot's Handbook of Aeronautical Knowledge*
FAA-H-8083-30	*Aviation Maintenance Technician Handbook—General*
FAA-H-8083-31	*Aviation Maintenance Technician Handbook—Airframe*
FAA-H-8083-32	*Aviation Maintenance Technician Handbook—Powerplant*
FAA-S-ACS-1	*Aviation Mechanic General, Airframe, and Powerplant Airman Certification Standards*
POH/AFM	*Pilot's Operating Handbook/FAA-Approved Airplane Flight Manual*

Be sure to use the latest versions of these references when reviewing for the test. Most of these documents are available on the FAA's website (www.faa.gov). Additionally, some of these resources (FAA handbooks, ACS, and Code of Federal Regulations) are reprinted by ASA (asa2fly.com) and are available from aviation retailers worldwide.

The Oral and Practical Tests

Prerequisites

All applicants must have met the prescribed experience requirements as stated in 14 CFR §65.77. In addition, all applicants must provide:

1. Proof of having unexpired passing credit for the Aviation Mechanic General (AMG) knowledge test by presenting an Airman Knowledge Test Report (except when properly authorized under the provisions of 14 CFR §65.80 to take the practical tests before the airman knowledge tests).
2. Identification with a photograph and signature.

Test Standards

The examiner will download an oral and practical examination that is generated for each applicant that reflects the knowledge, risk management, and skill elements for each relevant subject area as detailed in the Airman Certification Standards (ACS).

The ACS consists of three *Sections*: General, Airframe, and Powerplant. *Subjects* are the areas in which aviation mechanic applicants must have knowledge and demonstrate skill. The *Objective* of each subject states what the applicant should know, consider, and do, as appropriate.

The knowledge, skill, and risk management *Elements* are the items that should be performed or answered according to standards acceptable to the FAA.

References identify the publication(s) that describe the task. *(Information contained in manufacturer and/or FAA-approved data always takes precedence over textbook referenced data.)*

Element codes in the ACS are divided into four components. For example,

AM.I.A.K1:

AM = ACS (Aviation Mechanic)

I = Section (General)

A = Subject (Fundamentals of Electricity and Electronics)

K1 = Knowledge Element (Electron theory [conventional flow vs. electron flow.])

Knowledge test questions are linked to the ACS codes. The Airman Knowledge Test Report (AKTR) lists an ACS code that correlates to a specific subject element for a given section and subject. This will allow remedial instruction and re-testing to be specific and based on explicit learning criteria. The FAA encourages applicants and instructors to use the ACS when preparing for tests.

The following terms apply to each element:

- *Inspect* means to examine (with or without inspection enhancing tools/equipment).
- *Check* means to verify proper operation.
- *Troubleshoot* means to analyze and identify malfunctions.
- *Service* means to perform functions that assure continued operation.
- *Repair* means to correct a defective condition and repair of an airframe or powerplant system including component replacement and adjustment.
- *Overhaul* means to disassemble, clean, inspect, repair as necessary, and reassemble.

The applicant should be well prepared in *all* knowledge, skill, and risk management elements included in the standards.

Satisfactory performance to meet the requirements for certification is based on the applicant's ability to:

1. Show basic knowledge.
2. Demonstrate basic mechanic skills.
3. Perform the skill elements within the standards of the reference materials.

The practical test is passed if, in the judgment of the examiner, the applicant demonstrates the prescribed level of proficiency on the assigned elements in each subject area. Each practical examination item must be performed, at a minimum, to the performance level in the airman certification standards.

If, in the judgment of the examiner, the applicant does not meet the standards of any subject element performed, the associated subject is failed and therefore, the practical test is failed.

Typical areas of unsatisfactory performance and grounds for disqualification are:

1. Any action or lack of action by the applicant that requires corrective intervention by the examiner for reasons of safety.
2. Failure to follow recommended maintenance practices and/or reference material while performing projects.
3. Exceeding tolerances stated in the reference material.
4. Failure to recognize improper procedures.
5. The inability to perform to a return-to-service standard, where applicable.
6. Inadequate knowledge in any of the subject areas.

When an applicant fails a test, the examiner will record the applicant's unsatisfactory performance and elements not completed in terms of subjects appropriate to the practical test conducted.

The General Oral and Practical Tests

There are 12 subject areas that are tested on the General Oral and Practical Exams.

For each subject area, this guide provides typical oral questions and succinct answers for the knowledge and risk management elements and presents the skills that applicants must understand and demonstrate.

I. General

A. Fundamentals of Electricity and Electronics
B. Aircraft Drawings
C. Weight and Balance
D. Fluid Lines and Fittings
E. Aircraft Materials, Hardware, and Processes
F. Ground Operations and Servicing
G. Cleaning and Corrosion Control
H. Mathematics
I. Regulations, Maintenance Forms, Records, and Publications
J. Physics for Aviation
K. Inspection Concepts and Techniques
L. Human Factors

A. Fundamentals of Electricity and Electronics

References: AC 43.13-1; FAA-H-8083-30

Knowledge

1. **What is the difference between the conventional current flow theory and electron flow?** (AM.I.A.K1)

   Conventional current flow theory is a visualization of current as flowing from positive to negative. In actual practice, electrons in a circuit flow from negative to positive (electron theory).

2. **How can you find the polarity of an electromagnet?** (AM.I.A.K2)

   Hold the electromagnet in your left hand with your fingers encircling the coil in the direction electrons flow (from negative to positive). Your thumb will point to the north pole of the electromagnet.

3. **What constitutes a capacitor?** (AM.I.A.K3)

   Two conductors separated by an insulator.

4. **What is the purpose of a capacitor?** (AM.I.A.K3)

   It stores electrical energy in electrostatic fields.

5. **What is the basic unit of capacitance?** (AM.I.A.K3)

   The farad.

6. **Why should electrolytic capacitors not be used in an AC circuit?** (AM.I.A.K3)

   They are polarized. An electrolytic capacitor will pass current of one polarity, but will block current of the opposite polarity.

7. **What is meant by inductance?** (AM.I.A.K4)

   The ability to store electrical energy in electromagnetic fields.

8. **What is the basic unit of inductance?** (AM.I.A.K4)

   The henry.

9. **Given the inductance and the frequency of the AC in a circuit, how do you compute the inductive reactance caused by the coil?** (AM.I.A.K4)

   $X_L = 2\pi fL$

   Where: X_L = inductive reactance in ohms
   f = frequency in cycles per second
   π = 3.1416
   L = inductance

   For example, in an AC series circuit in which the inductance is 146 mH (millihenry) and the voltage is 110 volts at a frequency of 60 cps, the inductive reactance is determined by the following method:

   $X_L = 2\pi \times f \times L$
   $X_L = 6.28 \times 60 \times 0.146$
   $X_L = 55$ ohm

10. **Does a capacitor in an AC circuit cause the current to lead or lag the voltage?** (AM.I.A.K5)

    It causes the current to lead the voltage.

11. **Does an inductor in an AC circuit cause the current to lead or lag the voltage?** (AM.I.A.K5)

    It causes the current to lag behind the voltage.

12. **What is the basic unit of power in a DC circuit?** (AM.I.A.K6)

    The watt.

13. **What happens to the current in a DC circuit if the voltage is increased but the resistance remains the same?** (AM.I.A.K6)

    It increases.

14. **What are five sources of electrical energy?** (AM.I.A.K7)

    Magnetism, chemical energy, light, heat, and pressure.

15. **Which law of electricity is the most important for an aircraft mechanic to know?** (AM.I.A.K7)

    Ohm's law.

16. **What are the elements of Ohm's law?**

    Voltage E, current I, and resistance R (volts, amps, and ohms).

17. **What is the name of the law that describes the relationship in an electrical circuit of voltage, current, and resistance?** (AM.I.A.K7a)

    Ohm's law.

18. **What is the basic equation of Ohm's law?** (AM.I.A.K7a)

    $E = I \times R$

19. **What is Kirchhoff's voltage law?** (AM.I.A.K7b)

    Kirchhoff's voltage law states that the algebraic sum of all voltages around a closed path or loop is zero. Another way of saying it is that the sum of all the voltage drops equals the total source voltage.

20. **What is the formula for determining power?** (AM.I.A.K7c)

    General power formula: $P = I \times E$
    Where: P = Power, I = Current, E = Volts

21. **What is Faraday's law?** (AM.I.A.K7d)

    Faraday's law, or the law of electromagnetic induction, states that the induced EMF or electromagnetic force in a closed loop of wire is proportional to the rate of change of the magnetic flux through a coil of wire.

22. **What is Lenz's law?** (AM.I.A.K7e)

    Lenz's law states that the EMF induced in an electric circuit always acts in such a direction that the current it drives around a closed circuit produces a magnetic field, which opposes the change in magnetic flux.

23. **What is the right-hand motor rule?** (AM.I.A.K7f)
When the index finger of the right hand is pointed in the direction of the magnetic field and the second finger in the direction of current flow, the thumb indicates the direction the current carrying wire moves.

24. **What is meant by a kilowatt?** (AM.I.A.K8)
1,000 watts.

25. **What is meant by a megawatt?** (AM.I.A.K8)
1,000,000 watts.

26. **What instrument is used to measure current flow?** (AM.I.A.K8)
An ammeter.

27. **What is a megohm?** (AM.I.A.K8)
One million (1,000,000) ohms.

28. **What instrument is used to measure electrical resistance?** (AM.I.A.K8)
An ohmmeter.

29. **When measuring resistance of a component with an ohmmeter, should the circuit be energized?** (AM.I.A.K8)
No, there should be no power on the circuit.

30. **What instrument is used to measure continuity in an electrical circuit?** (AM.I.A.K8)
An ohmmeter.

31. **In what units is battery capacity expressed?** (AM.I.A.K8)
In ampere-hours.

32. **What instrument is used to measure the specific gravity of the electrolyte in a lead-acid battery?** (AM.I.A.K8)
A hydrometer.

33. **What is voltage?** (AM.I.A.K9)
Electrical pressure.

34. **What is the basic unit of voltage?** (AM.I.A.K9)
The volt.

35. **What instrument is used to measure voltage?** (AM.I.A.K9)
A voltmeter.

36. **To measure voltage, is a voltmeter placed in series or in parallel with the source of voltage?** (AM.I.A.K9)
In parallel.

37. **What is the voltage across each resistor connected in parallel across a 12-volt battery?** (AM.I.A.K9)
12 volts.

38. What is the open-circuit voltage of a lead-acid cell? (AM.I.A.K9)

2.1 volts.

39. How is output voltage regulated in an AC alternator? (AM.I.A.K9a)

Voltage regulation is accomplished by varying the strength of the AC exciter fields (field voltage).

40. What is electrical current? (AM.I.A.K10)

The flow of electrons in a circuit.

41. What is the basic unit of current flow? (AM.I.A.K10)

The ampere (amp).

42. What part of an amp is a milliamp? (AM.I.A.K10)

One thousandth (0.001) of an amp.

43. What two things happen when current flows through a conductor? (AM.I.A.K10)

Heat is generated and a magnetic field surrounds the conductor.

44. To measure current through a component, is an ammeter placed in parallel or in series with the component? (AM.I.A.K10)

In series.

45. What formula is used to find current when voltage and resistance are known? (AM.I.A.K10)

Current = Voltage divided by Resistance.

$$I = \frac{E}{R}$$

46. What is meant by resistance in an electrical circuit? (AM.I.A.K11)

The opposition to the flow of electrons.

47. What is the basic unit of electrical resistance? (AM.I.A.K11)

The ohm.

48. What four things affect the resistance of an electrical conductor? (AM.I.A.K11)

The material, the cross-sectional area, the length, and the temperature.

49. How can you tell the resistance of a composition resistor? (AM.I.A.K11)

By a series of colored bands around one end of the resistor.

50. What formula is used to find resistance when voltage and current are known? (AM.I.A.K11)

Resistance = Voltage divided by Current

$$R = \frac{E}{I}$$

51. What is meant by impedance? (AM.I.A.K11a)

The total opposition to the flow of alternating current. It is the vector sum of resistance, capacitive reactance, and inductive reactance.

52. In what units is impedance measured? (AM.I.A.K11a)

In ohms.

53. How is resistance calculated in a series circuit? (AM.I.A.K11b)

For resistors in a series configuration, the total resistance of the circuit is equal to the sum of the individual resistors.

54. How is resistance calculated in a parallel circuit? (AM.I.A.K11c)

The formula for the total parallel resistance is as follows:

$$\frac{1}{R_T} = \frac{1}{R_1} + \frac{1}{R_2} + \frac{1}{R_3} + \ldots \frac{1}{R_N}$$

If the reciprocal of both sides is taken, then the general formula for the total parallel resistance is:

$$R_T = \frac{1}{\frac{1}{R_1} + \frac{1}{R_2} + \frac{1}{R_3} + \ldots \frac{1}{R_N}}$$

55. What is the total resistance of three 12-ohm resistors connected in series? (AM.I.A.K11d)

36 ohms.

56. What is the total resistance of three 12-ohm resistors connected in parallel? (AM.I.A.K11d)

4 ohms.

57. What is the formula for power in a DC circuit? (AM.I.A.K12)

Power = Voltage times current ($P = E \times I$)

58. What is the relationship between mechanical and electrical power? (AM.I.A.K12)

1 horsepower = 746 watts

59. What is meant by true power in an AC circuit? (AM.I.A.K12)

The product of the circuit voltage and the current that is in phase with this voltage.

60. In what units is true power expressed? (AM.I.A.K12)

In watts.

61. What is meant by apparent power in an AC circuit? (AM.I.A.K12)

The product of the circuit voltage and the circuit current, without reference to phase angle.

62. In what units is apparent power expressed? (AM.I.A.K12)

In volt-amps.

63. What is meant by reactive power in an AC circuit? (AM.I.A.K12)

The power consumed in the inductive and capacitive reactances in an AC circuit. Reactive power is also called wattless power.

64. In what units is reactive power expressed? (AM.I.A.K12)

In volt-amps reactive (VAR), or kilovolt-amps reactive (KVAR).

65. **What is meant by power factor in an AC circuit?** (AM.I.A.K12)

The ratio of true power to apparent power. It is also the ratio of circuit resistance to circuit impedance.

66. **How many watts of power are consumed by a 1/5 horsepower, 24-volt DC motor that is 75% efficient?** (AM.I.A.K12)

Power efficiency is expressed as

$$\text{Efficiency} = \frac{\text{Power Out}}{\text{Power In}}$$

Power consumed is the same as Power In, and 1 horsepower (hp) equals 746 watts. Therefore

$$\text{Power In} = \frac{\text{Power Out}}{\text{Efficiency}} = \frac{\frac{1}{5}\text{hp} \times 746\frac{\text{Watts}}{\text{hp}}}{0.75} = \frac{0.2\text{ hp (746 W/hp)}}{0.75} = 199\text{ Watts}$$

67. **How much current flows through each of three resistors connected in series if the total current is 3 amps?** (AM.I.A.K13)

3 amps.

68. **How much current flows through each of three equal resistors connected in parallel if the total current is 3 amps?** (AM.I.A.K14)

1 amp.

69. **What is meant by the capacity rating of a lead-acid battery?** (AM.I.A.K15)

The number of hours a battery can supply a given current flow.

70. **What electrolyte is used in a lead-acid battery?** (AM.I.A.K15)

A mixture of sulfuric acid and water.

71. **Does the specific gravity of the electrolyte in a lead-acid battery increase or decrease as the battery becomes discharged?** (AM.I.A.K15)

It decreases.

72. **What is the specific gravity of a fully charged lead-acid battery?** (AM.I.A.K15)

Between 1.275 and 1.300.

73. **How many cells are there in a 24-volt lead-acid battery?** (AM.I.A.K15)

12

74. **What is the range of temperatures of the electrolyte in a lead-acid battery that does not require a correction when measuring its specific gravity?** (AM.I.A.K15)

Between 70°F and 90°F.

75. **How do you treat a lead-acid battery compartment to protect it from corrosion?** (AM.I.A.K15)

Paint it with an asphaltic (tar-base) paint or with a polyurethane enamel.

76. **What is used to neutralize spilled electrolyte from a lead-acid battery?** (AM.I.A.K15)

A solution of bicarbonate of soda and water.

77. How high should the electrolyte level be in a properly serviced lead-acid battery? (AM.I.A.K15)

Only up to the level of the indicator in the cell.

78. Why is the closed-circuit voltage of a lead-acid battery lower than its open-circuit voltage? (AM.I.A.K15)

Voltage is dropped across the internal resistance of the battery.

79. What gases are released when a lead-acid battery is being charged? (AM.I.A.K15)

Hydrogen and oxygen.

80. What is the electrolyte used in a nickel-cadmium battery? (AM.I.A.K15)

Potassium hydroxide and water.

81. Why is a hydrometer not used to measure the state of charge of a nickel-cadmium battery? (AM.I.A.K15)

The electrolyte of a nickel-cadmium battery does not enter into the chemical changes that occur when the battery is charged or discharged. Its specific gravity does not change appreciably.

82. Is the electrolyte level of a nickel-cadmium battery lowest when the battery is fully charged or fully discharged? (AM.I.A.K15)

Fully discharged.

83. What is a result of cell imbalance in a nickel-cadmium battery? (AM.I.A.K15)

The low internal resistance allows current to flow between the unbalanced cells and generate heat.

84. What is a thermal runaway? (AM.I.A.K15)

The large current flow allowed by the low internal resistance causes the cells to produce more heat than they can dissipate. The heat further lowers the internal resistance so more current can flow; this continues until the battery destroys itself.

85. How may thermal runaway be prevented? (AM.I.A.K15)

By carefully monitoring the temperature of the cells and controlling the charging current to prevent an excess of current flowing into the battery.

86. How is it possible to know when a nickel-cadmium battery is fully charged? (AM.I.A.K15)

Completely discharge the battery and give it a constant-current charge to 140% of its ampere-hour capacity.

87. What is used to neutralize spilled electrolyte from a nickel-cadmium battery? (AM.I.A.K15)

A solution of boric acid and water.

88. What is a transformer, and how does it work? (AM.I.A.K16)

A transformer is a device that changes electrical energy in an alternating current (AC) circuit of a given voltage into electrical energy at a different voltage level. It consists of two coils that are not electrically connected but arranged so that the magnetic field surrounding one coil cuts through the other coil.

89. What three things must an electric circuit contain? (AM.I.A.K17)

A source of electrical energy, a load to use the energy, and conductors to join the source and the load.

90. What is meant by continuity in an electrical circuit? (AM.I.A.K17)

The circuit is continuous (or complete) when electrons can flow from one terminal of the power source to the other.

91. What is the basic function of a switch? (AM.I.A.K18)

Switches control the current flow in most aircraft electrical circuits.

92. What is a relay? (AM.I.A.K18)

A relay is simply an electromechanical switch where a small amount of current can control a large amount of current.

93. What is the function of a current limiter? (AM.I.A.K19)

A current limiter is a type of heavy-duty fuse, commonly with ratings of 30 amps or greater. They are often used to protect a section of an electrical system, such as a single electrical bus.

94. What materials are fixed resistors made of? (AM.I.A.K20)

The most common fixed resistors are made of a carbon composition. Other types of fixed resistors include carbon film, metal-oxide, metal film, and metal glaze. Wire-wound resistors are typically used to control large amounts of current and have high power ratings.

95. What are the two main types of variable resistors? (AM.I.A.K20)

The two main types of variable resistors are the rheostat and the potentiometer.

96. How can you tell the resistance of a composition resistor? (AM.I.A.K20)

By a series of colored bands around one end of the resistor.

97. What happens to the current in a conductor if the length of the conductor is doubled with all other parameters unchanged? (AM.I.A.K21)

The resistance of the conductor would double and current flow would be decreased.

98. Explain the basic concept of digital logic, including RAM, ROM, NVRAM, logic gates, inverters, rectifiers, and flip flops. (AM.I.A.K22)

Transistors are used in digital electronics to construct circuits that act as digital logic gates. The purpose and task of a device is achieved by manipulating electric signals through the logic gates. Thousands, and even millions, of tiny transistors can be placed on a chip to create the digital logic landscape through which a component's signals are processed.

(continued)

a. *RAM*—Random access memory that is used for temporary storage in a computer system.
b. *ROM*—Read-only memory is a type of nonvolatile memory in a computer that is not lost when power to the computer is lost.
c. *NVRAM*—Non-volatile random access memory is random access memory that is retained, even when power is lost.
d. *Logic gates*—Logic gates provide a set output signal based on one or more input signals. The input signal is either voltage (referred to as True, Logic 1, or Yes), or no voltage (referred to as False, Logic 0, or No). The basic logic gates are AND, OR and NOT.
e. *Inverter*—When used as a logic function, converts a 1 to a 0, or a 0 to a 1.
f. *Rectifier*—A rectifier converts AC voltage to DC voltage.
g. *Flip-flop*—An electrical circuit that can store a single binary signal, either a 1 or 0. The binary state can change based on the input signal.

99. What are binary numbers? (AM.I.A.K23)

The binary number system has only two digits: 0 and 1. It is also known as a base-2 numbering system.

100. What is electrostatic discharge (or ESD)? (AM.I.A.K24)

Electrostatic discharge (ESD) is the discharge of static electricity that can build up on a component or a person's body. ESD is a frequent cause of damage to solid-state components and integrated circuits and can be prevented by wearing a grounding wrist strap when handling electrical components.

101. What type of drawing depicts electrical components with respect to each other within a circuit? (AM.I.A.K25)

An electrical schematic.

102. What are three types of DC circuits with regard to the placement of the various circuit components? (AM.I.A.K26)

Series, parallel, and series-parallel.

103. What are the two general types of AC motors used in aircraft systems? (AM.I.A.K27)

Induction motors and synchronous motors.

104. What is the formula used to determine the speed of an AC motor? (AM.I.A.K27)

$$\text{RPM} = \frac{(120 \times \text{Frequency})}{\text{Number of Poles}}$$

105. In what types of applications is a series wound DC motor most often used? (AM.I.A.K27)

Applications that require high torque and low speed, such as a starter motor.

106. What are the two principal parts of a DC motor? (AM.I.A.K27)

The field assembly and an armature assembly.

Risk Management

1. **When a volt/ohm multimeter is set to measure current, what precautions should be taken?** (AM.I.A.R1)

   Never place the multimeter leads across a voltage drop, as this can damage the meter.

2. **What precautions should be taken in the maintenance shop where both lead-acid and nickel-cadmium batteries are serviced?** (AM.I.A.R2)

   The two types of batteries should be kept separate, and the tools used on one type should not be used on the other.

3. **What precaution should be taken when working on a high voltage strobe light circuit?** (AM.I.A.R3)

   Disconnect the power from the strobe power supply and let it stand for 5–10 minutes to allow the capacitors to bleed off.

4. **What are some precautions that should be taken when working around an aircraft battery installed in an aircraft?** (AM.I.A.R4)

   a. When disconnecting a battery, always disconnect the negative terminal first.
   b. When installing a battery, always connect the negative terminal last.
   c. When the battery is installed, care should be taken to never allow tools or equipment to short the positive terminal to ground.

Skills

1. **Using a capacitor tester and capacitor furnished by the examiner, determine the capacity of a capacitor and whether or not it is serviceable.** (AM.I.A.S1)

2. **Find the capacitive reactance in an AC circuit for the values of capacitance and frequency specified by the examiner.** (AM.I.A.S1)

3. **Find the inductive reactance in an AC circuit for the values of inductance and frequency specified by the examiner.** (AM.I.A.S1)

4. **Install wires into an electrical connector assigned by the examiner and test for continuity.** (AM.I.A.S1)

5. **Compute the amps of current drawn by a 1,000-watt landing light in a 24-volt DC electrical system.** (AM.I.A.S2)

6. **Find the total number of watts dissipated by two lamps wired in parallel in a 12-volt circuit, if one lamp requires 3 amps and the other 1.5 amp.** (AM.I.A.S2)

7. **Find the current through each resistor, the voltage drop across each resistor, and the power dissipated by each resistor in this circuit. Resistor values will be provided by the examiner.** (AM.I.A.S2)

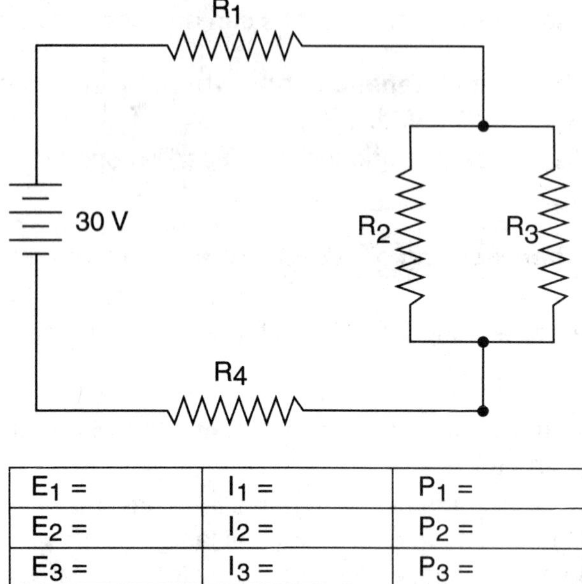

E_1 =	I_1 =	P_1 =
E_2 =	I_2 =	P_2 =
E_3 =	I_3 =	P_3 =
E_4 =	I_4 =	P_4 =

8. **Using a multimeter and an energized electrical circuit, measure the voltage drop across a color-coded resistor. Compute the current flow through it and the amount of power dissipated in it.** (AM.I.A.S2)

9. **Measure the DC current through a component specified by the examiner.** (AM.I.A.S3)

10. **Measure the resistance of a component specified by the examiner and determine whether or not it is within the limits specified in an appropriate service manual.** (AM.I.A.S4)

11. **Using an ohmmeter, determine if a supplied switch is operating correctly.** (AM.I.A.S5)

12. **Using a multimeter, demonstrate to the examiner how to determine if a fuse is blown, or if a circuit breaker is working correctly.** (AM.I.A.S6)

13. **Using an electrical schematic circuit diagram furnished by the examiner, trace the flow of current from a battery through a series of switches and relays to a component.** (AM.I.A.S7)

14. **Using a logic diagram of an electronic circuit, explain to the examiner the function of AND, OR, NOT, and NOR gates.** (AM.I.A.S7)

15. **Demonstrate to the examiner the way to trace a procedure using a logic flow chart.** (AM.I.A.S7)

16. **Use a voltmeter to troubleshoot a supplied circuit.** (AM.I.A.S8)

17. **Explain to the examiner the various symbols used on an electrical system schematic diagram.** (AM.I.A.S9)

18. **Explain to the examiner how to test a circuit for short-circuit and open-circuit conditions.** (AM.I.A.S10)

19. **Measure the voltage drop across a resistor in an energized DC circuit specified by the examiner.** (AM.I.A.S11)

20. **Demonstrate to the examiner the correct way to check an electrical circuit for continuity.** (AM.I.A.S12)

21. **Visually inspect a battery and battery compartment for defects and provide a list of discrepancies.** (AM.I.A.S13)

22. **Check the battery for cell imbalance.** (AM.I.A.S13)

23. **Check electrolyte level and service as necessary.** (AM.I.A.S14)

24. **Check the state-of-charge of a lead-acid battery using a hydrometer. Apply the appropriate correction for temperature.** (AM.I.A.S14)

25. **Properly remove a lead-acid battery from an aircraft. Describe the proper way to clean the outside of the battery and reinstall it, correctly connecting the battery cables.** (AM.I.A.S14)

26. **Properly connect two batteries on a charger for a constant-voltage charge.** (AM.I.A.S14)

27. **Properly connect two batteries on a charger for a constant-current charge.** (AM.I.A.S14)

28. **Remove a nickel-cadmium battery from an aircraft. Describe to the examiner the correct way to clean the battery and battery compartment.** (AM.I.A.S14)

29. **Perform a deep-cycling discharge and recharge of the battery on a constant-current battery charger.** (AM.I.A.S14)

30. **Properly reinstall the battery in the aircraft and connect the battery cables.** (AM.I.A.S14)

B. Aircraft Drawings

References: AC 43.13-1; FAA-H-8083-30

Knowledge

1. **How many views can there be on an orthographic projection?** (AM.I.B.K1)
   Six.

2. **How many views are used to show most objects in an aircraft drawing?** (AM.I.B.K1)
   Three.

3. **What information is given in the title block of an aircraft drawing?** (AM.I.B.K1)
   The name and address of the company who made the part, the name of the part, the scale of the drawing, the name of the draftsman, the name of the engineer approving the part, and the number of the part (the drawing number).

4. **Where is the title block normally located on an aircraft drawing?** (AM.I.B.K1)
   In the lower right-hand corner of the drawing.

5. **How can you know that the aircraft drawing you are using is the most current version of the drawing?** (AM.I.B.K1)
   By the number in the revision block, and by the log of the most recent drawings.

6. **What is a detail drawing?** (AM.I.B.K1)
   A drawing that includes all the information needed to fabricate the part.

7. **What is an assembly drawing?** (AM.I.B.K1)
   A drawing that shows all the components of a part in an exploded form. A parts list is included with an assembly drawing.

8. **What is an installation drawing?** (AM.I.B.K1)
   A drawing that shows the location of the parts and assemblies on the complete aircraft.

9. **What type of drawing is most helpful in troubleshooting a system?** (AM.I.B.K1)
   A schematic diagram.

10. **What is a block diagram?** (AM.I.B.K1)
    A drawing that shows the various functions of a system by a series of blocks. These blocks do not include any detail, but instead indicate what happens in each block.

11. **How are dimensions shown on an aircraft drawing?** (AM.I.B.K1)
    By numbers shown in the break of a dimension line.

12. **What is the purpose of a center line on an aircraft drawing?** (AM.I.B.K1)
    It divides a part into symmetrical halves.

13. **What is the purpose of a sketch of a repair?** (AM.I.B.K2)

    It shows a specific bit of information and includes the minimum amount of detail needed to manufacture the part.

14. **Where would an aviation mechanic find aircraft electrical schematics?** (AM.I.B.K3)

    In the aircraft maintenance manual or the aircraft wiring diagram manual.

15. **What is a fuselage station number?** (AM.I.B.K4)

    The distance in inches from the datum, measured along the longitudinal axis of the fuselage.

16. **What is a butt line?** (AM.I.B.K4)

    A reference line to the right or left of the center line of the aircraft.

Risk Management

1. **Why is it important to understand and follow tolerances on aircraft drawings?** (AM.I.B.R1)

   Tolerances on a drawing control how parts or components fit together. Incorrectly sized parts could cause damage to the aircraft structure or systems.

2. **What is the risk of not following the exact specification for the design of alterations or repairs?** (AM.I.B.R2)

   Not following the specifications will cause the alteration or repair to be outside of the approved parameters, potentially leading to an unairworthy and unsafe condition.

3. **Explain how an aviation mechanic can mitigate applicability risk when using a drawing or schematic.** (AM.I.B.R3)

   Always confirm that the drawing or schematic applies to the make, model, serial number, and configuration of the aircraft that it is being used for.

4. **What risks are associated with not using the most current version of a drawing, or not verifying that the drawing applies to the aircraft?** (AM.I.B.R4)

   Not using the most current revision of a drawing could result in missing important updates.

Skills

1. **What are the four steps in making a sketch?** (AM.I.B.S1)

   1. Block in the space and basic shape used for the sketch.
   2. Add details to the basic block.
   3. Darken lines that are to show up as visible lines in the finished sketch.
   4. Add dimensions and any other information that will make the sketch more usable.

2. **Draw a sketch of a repair or alteration specified by the examiner. Use the necessary drafting tools, such as dividers, compass, ruler, T-square, etc.** (AM.I.B.S1)

3. **Draw a three-dimensional sketch of a component shown on a three-view drawing.** (AM.I.B.S1)

4. **Identify each of these lines that are commonly used on an aircraft drawing.** (AM.I.B.S2)

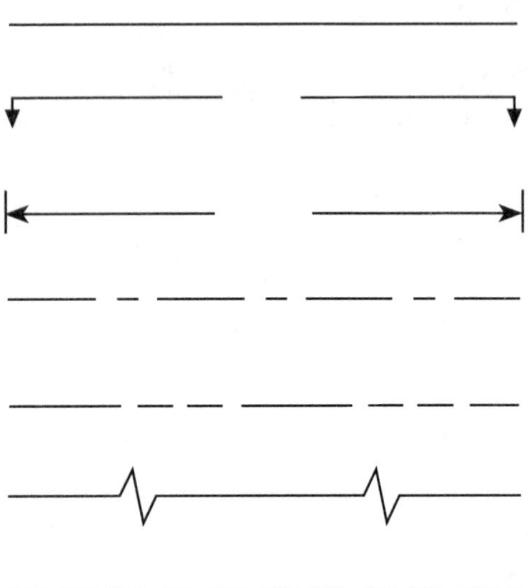

5. **Using the schematic diagram of an aircraft electrical system specified by the examiner, answer a series of questions regarding hypothetical troubleshooting situations.** (AM.I.B.S2)

6. **Explain to the examiner what information a selected installation diagram or schematic provides.** (AM.I.B.S2)

7. **Identify the lines and symbols on an aircraft drawing specified by the examiner.** (AM.I.B.S2)

8. **On an aircraft drawing specified by the examiner, locate the following items:**
   - **title block**
   - **zone numbers**
   - **notes**
   - **bill of materials**
   - **change number**

   (AM.I.B.S2)

9. **Demonstrate to the examiner the correct way to dimension a component.** (AM.I.B.S3)

10. **Explain to the examiner what the changes were on the latest revision of the supplied drawing.** (AM.I.B.S4)

11. **Determine the material requirements required to fabricate a part from a drawing.** (AM.I.B.S5)

12. **Demonstrate to the examiner your ability to assemble a component using an assembly drawing.** (AM.I.B.S6)

13. **Using a performance chart for a specific aircraft engine, find the brake horsepower developed when the RPM and BMEP are known.** (AM.I.B.S6)

C. Weight and Balance

References: AC 43.13-1; FAA-H-8083-1; FAA-H-8083-30

Knowledge

1. **What is meant by tare weight?** (AM.I.C.K1)

   The weight of any chocks and other items that are used to hold the aircraft on the scales.

2. **What is included in the empty weight of an aircraft?** (AM.I.C.K1)

   The weight of the airframe, engines, and all items of operating equipment that have fixed locations and are permanently installed in the aircraft. Empty weight includes optional and special equipment, fixed ballast, full reservoirs of hydraulic fluid and engine lubricating oil, and unusable fuel.

3. **What equipment must be installed in an aircraft when it is weighed to find its empty weight center of gravity?** (AM.I.C.K1)

   All the equipment listed in the Aircraft Equipment List as "required equipment" or as equipment that is permanently installed.

4. **What is meant by permanent ballast for an aircraft?** (AM.I.C.K1)

   Weight that is permanently installed in an aircraft to bring the empty weight center of gravity into allowable limits.

5. **What is meant by the maximum zero fuel weight of an aircraft?** (AM.I.C.K1)

   The maximum permissible weight of a loaded aircraft (passengers, crew, cargo, etc.), less its fuel.

6. **What is meant by undrainable fuel?** (AM.I.C.K1)

   The fuel that is left in the tank, lines and components when the aircraft is placed in level flight position and the fuel drained at the main fuel strainer. This is also called residual fuel.

7. **What is meant by the datum that is used for weight and balance computations?** (AM.I.C.K1)

   It is a readily identified reference chosen by the aircraft manufacturer from which all longitudinal locations on the aircraft are referenced.

8. **What is meant by moment in the computation of weight and balance?** (AM.I.C.K1)
   A force that tends to cause rotation. It is the product of the weight of an object in pounds and the distance of the object from the datum in inches.

9. **What is meant by the arm of an item installed in an aircraft?** (AM.I.C.K1)
   The distance, in inches, between the center of gravity of the item and the datum.

10. **What is meant by a moment index?** (AM.I.C.K1)
    A moment divided by a constant such as 10, 100, or 1,000.

11. **What is meant by the center of gravity range?** (AM.I.C.K1)
    The distance in inches between the forward allowable center of gravity and the rearward allowable center of gravity.

12. **What is a loading envelope?** (AM.I.C.K1)
    The enclosed area on a graph of an airplane's loaded weight and CG location. If lines drawn from the weight and CG cross within this envelope the airplane is properly loaded.

13. **What is meant by permanent ballast for an aircraft?** (AM.I.C.K1)
    Weight that is permanently installed in an aircraft to bring the empty weight center of gravity into allowable limits.

14. **When should an aircraft be reweighed?** (AM.I.C.K2)
    After an extensive repair or alteration that could change its weight or center of gravity.

15. **Why is it necessary to consider the category under which an aircraft is licensed when computing its weight and balance?** (AM.I.C.K3)
    Different categories under which an aircraft can be licensed have different maximum gross weights and different center of gravity ranges.

16. **Where do you find the leveling means that are specified for a particular aircraft?** (AM.I.C.K3)
    In the Type Certificate Data Sheets for the aircraft.

17. **What must be done with tare weight when an aircraft is weighed?** (AM.I.C.K3)
    It must be subtracted from the scale readings to find the weight of the aircraft.

18. **Why are the distances of all the items installed in an aircraft measured from the datum when computing weight and balance?** (AM.I.C.K3)
    This makes it possible to find the point about which the aircraft would balance (the center of gravity).

19. **Where is the maximum allowable gross weight of an aircraft found?** (AM.I.C.K3)
    In the Type Certificate Data Sheets for the aircraft.

20. **Is there a Federal Regulation requiring that all private aircraft be reweighed periodically?** (AM.I.C.K3)
    No.

21. What must be done to find the empty weight of an aircraft if it has been weighed with fuel in its tanks? (AM.I.C.K3)

The weight of the fuel and its moment must be subtracted from the weight and moment of the aircraft as it was weighed.

22. How would you find the empty weight and empty weight center of gravity of an airplane if there are no weight and balance records available? (AM.I.C.K3)

The aircraft is weighed, and the empty weight center of gravity is computed. These values are recorded in new weight and balance records started for the aircraft.

23. Which has the more critical center of gravity range, an airplane or a helicopter? (AM.I.C.K4)

A helicopter.

24. Where is the center of gravity for most airplanes located in relation to the center of lift? (AM.I.C.K4)

The CG is normally ahead of the center of lift.

25. How do you find the moment of an item installed in an aircraft? (AM.I.C.K4)

Multiply the weight of the item in pounds by its distance from the datum in inches.

26. Where do you find the arm of an item installed in an aircraft? (AM.I.C.K4)

In the Type Certificate Data Sheets or the aircraft equipment list.

27. What is the significance of the empty weight center of gravity range of an aircraft? (AM.I.C.K5)

If the empty weight center of gravity falls within the EWCG range, the aircraft cannot be legally loaded in such a way that its loaded center of gravity will fall outside of the allowable loaded CG range. Not all aircraft have an EWCG range.

28. What are two reasons weight and balance control are important in an aircraft? (AM.I.C.K6)

For safety of flight and for most efficient operation of the aircraft.

29. What is meant by minimum fuel as is used in the computation of aircraft weight and balance? (AM.I.C.K7)

No more fuel than the quantity necessary for one-half hour of operation at rated maximum continuous power. It is the maximum amount of fuel used in weight and balance computations when low fuel may adversely affect the most critical balance conditions.

30. Why is the empty weight center of gravity range not given in the Type Certificate Data Sheets for some aircraft? (AM.I.C.K8)

The empty weight center of gravity range is given only for aircraft that cannot be legally loaded in such a way that their loaded center of gravity will fall outside of the allowable limits.

31. Where must a record be kept of the current empty weight and the current center of gravity of an aircraft? (AM.I.C.K8)

In the aircraft flight manual or weight and balance records required by 14 CFR §23.1583.

32. **What must be done with temporary ballast when weighing an aircraft?** (AM.I.C.K9)
    It must be removed from the aircraft.

33. **Where can you find instructions for jacking the complete aircraft for weighing, or to perform a landing gear retraction test?** (AM.I.C.K10)
    In the aircraft maintenance manual.

Risk Management

1. **Where can you find the instructions for safely jacking an aircraft?** (AM.I.C.R1)
   In the aircraft maintenance manual.

2. **To reduce the risks of an improperly recorded empty weight center of gravity, where would one find the correct procedures for weighing a specific aircraft?** (AM.I.C.R2)
   In the aircraft maintenance manual.

3. **Where can you find the procedures for using scales or load cells for weighing an aircraft?** (AM.I.C.R3)
   In the aircraft maintenance manual.

4. **What are the dangers of the incorrect use of scales when weighing an aircraft?** (AM.I.C.R3)
   The improper use of scales will cause the recording of incorrect aircraft weights that could cause the aircraft to be loaded outside of safe limits.

5. **What is the aerodynamic effect danger of a CG that is either forward or aft of CG limits?** (AM.I.C.R4)
   The aircraft may become uncontrollable, leading to an accident.

6. **What is the effect if an aircraft is loaded in excess of the maximum certified weight?** (AM.I.C.R5)
   Takeoff and climb performance will decrease and structural limits may be exceeded, all of which could cause an accident.

Skills

1. **Demonstrate to the examiner the correct way to level an aircraft for weighing.** (AM.I.C.S1)

2. **Locate the "leveling means" for an aircraft specified by the examiner.** (AM.I.C.S1)

3. **Explain to the examiner the way an airplane is prepared for weighing.** (AM.I.C.S1)

4. **Demonstrate to the examiner the way to use a aircraft maintenance manual to find the procedure used to jack the aircraft for weighing.** (AM.I.C.S1)

5. **Weigh an aircraft and record the scale readings, including proper handling of tare items.** (AM.I.C.S2)

6. **Calculate the minimum fuel for weight and balance calculations.** (AM.I.C.S2)

7. **Compute the center of gravity for a helicopter.** (AM.I.C.S2)

8. **Find the amount of ballast needed to bring this aircraft into its proper center of gravity range:**
   Aircraft as loaded weighs 4,954 pounds.
   Aircraft loaded center of gravity is +30.5 inches aft of the datum.
   Loaded center of gravity range is +32.0 to +42.1 inches aft of the datum.
   The ballast arm is +162 inches.
   (AM.I.C.S3)

9. **Check the aircraft weighing scales for calibration.** (AM.I.C.S4)

10. **Calculate the new weight and balance on an aircraft after the installation or removal of equipment specified by the examiner.** (AM.I.C.S5)

11. **Determine the distance between the forward and aft center of gravity limits for a helicopter.** (AM.I.C.S6)

12. **Revise the weight and balance records of an aircraft to reflect the changes made by an alteration specified by the examiner.** (AM.I.C.S7)

13. **Create a logbook entry for the completion of a new weight and balance based on reweighing the aircraft.** (AM.I.C.S7)

14. **Find the empty weight and the empty weight center of gravity of an airplane that has the following scale weights:**
    Left main wheel = 1,765 pounds, arm = +195.5 inches
    Right main wheel = 1,775 pounds, arm = +195.5 inches
    Nose wheel = 2,322 pounds, arm = +83.5 inches
    (AM.I.C.S8)

15. **Find the empty weight and empty-weight center of gravity of an aircraft specified by the examiner.** (AM.I.C.S8)

16. **Demonstrate to the examiner the way to calculate empty weight and CG in inches from the datum and as a percentage of the mean aerodynamic chord.** (AM.I.C.S8)

17. **Find the new empty weight and empty weight center of gravity for this aircraft after it has been altered by removing two seats and replacing them with a cabinet, one seat, and some radio gear.**
    Aircraft empty weight = 2,886 pounds
    Empty weight total moment = 107,865.78
    Each removed seat weighs 15 pounds, located at station 73.
    Installed cabinet weighs 97 pounds, installed at station 73.
    New seat weighs 20 pounds, installed at station 73.
    Radio gear weighs 30 pounds, installed at station 97.
    (AM.I.C.S8)

18. **Given the weight and balance manual for an aircraft, find the empty weight and EWCG of the aircraft.** (AM.I.C.S8)

19. **Calculate the moment of a specified item of equipment.** (AM.I.C.S9)

20. **Identify the tare items used when weighing an aircraft.** (AM.I.C.S10)

21. **Demonstrate to the examiner where the weight and balance records are in the supplied aircraft maintenance records.** (AM.I.C.S11)

22. **Locate the datum of an aircraft specified by the examiner.** (AM.I.C.S12)

23. **Locate the baggage compartment placard requirements for an assigned aircraft.** (AM.I.C.S13)

24. **Demonstrate to the examiner where to find the loading limitations on an assigned aircraft.** (AM.I.C.S13)

25. **Revise an aircraft equipment list after an equipment change.** (AM.I.C.S14)

26. **Calculate the change needed to correct an out-of-balance condition.** (AM.I.C.S15)

27. **Demonstrate to the examiner how to determine the approved CG range on a given aircraft based on aircraft specifications and Type Certificate Data Sheets (TCDS).** (AM.I.C.S16)

28. **Create a maintenance record entry for a change to aircraft weight and balance.** (AM.I.C.S17)

D. Fluid Lines and Fittings

References: AC 43.13-1; FAA-H-8083-30

Knowledge

1. **Of what material are most low-pressure rigid fluid lines made?** (AM.I.D.K1)
   1100-1/2 hard or 3003-1/2 hard aluminum alloy tubing.

2. **Is the size of a rigid fluid line determined by its inside or its outside diameter?** (AM.I.D.K1)
   By its outside diameter.

3. **What is the difference between the flare angle for aircraft and automotive fittings?** (AM.I.D.K1)
   Aircraft fittings have a 37° flare angle and automotive fittings use 45°.

4. **Where are quick-disconnect fluid line couplings normally used in an aircraft hydraulic system?** (AM.I.D.K1)
   In the lines that connect the engine-driven pump into the hydraulic system.

5. **Is the size of a flexible hose determined by its inside or its outside diameter?** (AM.I.D.K1)
   By its inside diameter.

6. **Of what material should rigid fluid lines be made that carry high-pressure (3,000 psi or greater) hydraulic fluid?** (AM.I.D.K2)
   Annealed or 1/4-hard corrosion-resistant steel.

7. **What is the principal advantage of Teflon hose for use in an aircraft hydraulic system?** (AM.I.D.K2)
   Teflon hose retains its high strength under conditions of high temperature.

8. **How tight should an MS flareless fitting be tightened?** (AM.I.D.K3)
   Tighten the fitting by hand until it is snug, and then turn it with a wrench for 1/6-turn to 1/3-turn. Never turn it more than 1/3-turn with a wrench.

9. **What damage can be caused by overtightening an MS flareless fitting?** (AM.I.D.K3)
   Overtightening drives the cutting edge of the sleeve deep into the tube and weakens it.

10. **What kind of rigid tubing can be flared with a double flare?** (AM.I.D.K3)
    5052-O and 6061-T aluminum alloy tubing in sizes from 1/8-inch to 3/8-inch OD.

11. **What should be done to the end of a tube that is to be flared?** (AM.I.D.K3)
    The cut end should be polished to remove any sharp edges that could cause the tubing to crack.

12. **What is the minimum amount of slack that must be left when a flexible hose is installed in an aircraft hydraulic system?** (AM.I.D.K4)
    The hose should be at least 5% longer than the distance between the fittings. This extra length provides the needed slack.

13. **How much pressure is used to proof-test a flexible hose assembly?** (AM.I.D.K4)
    This varies with the hose, but it is generally about two times the recommended operating pressure for the hose.

14. **Why is it important to use a torque wrench when securing fluid hose and line fittings?** (AM.I.D.K5)
    If the connection is too loose, it may leak. If the connection is too tight, it can cause the fitting to become damaged and fail.

15. **What is the purpose of using torque seal on a critical fitting after installation?** (AM.I.D.K6)
    If a visual inspection reveals that the torque seal has broken, it is an indication that the fitting may have loosened.

Risk Management

1. **Why is it important to understand system configurations during aircraft maintenance?** (AM.I.D.R1)
   A lack of understanding of a system could lead to fluid lines being reversed or incorrectly installed, leading to a potentially hazardous situation.

2. **What safety equipment should be considered when working on fluid lines?** (AM.I.D.R2)
   Depending on the types of fluid, safety equipment may include gloves and clothing protection, eye protection, spill containment materials, respirator, and fire extinguisher.

3. **How can an aviation mechanic determine what fluids are contained in the hoses and fluid lines that are encountered when working on an aircraft?** (AM.I.D.R3)
   Fluid lines and hoses are labeled at regular intervals to identify the type of fluid. Additionally, the aircraft maintenance manual provides information on each aircraft system.

4. **What risk mitigation action should occur prior to working on any pressurized fluid system?** (AM.I.D.R4)
   The system must be depressurized prior to beginning any work on the system.

5. **What is the function of the lay line (the identification stripe) that runs the length of a flexible hose?** (AM.I.D.R5)
   To show the mechanic whether or not the line was twisted when it was installed. The line should be straight, not spiraled. A twisted hose may move from its intended position and chafe or interfere with other systems.

6. **What is the danger of a loosened fitting or a hose that has moved out of position?** (AM.I.D.R6)
   A loosened fitting may leak or allow a hose to move out of position. A hose that has moved out of position may chafe on adjacent structures and begin to leak, or it may cause interference and damage to other systems.

7. **What precaution must be used when applying torque to a fluid line?** (AM.I.D.R7)
   The fitting that the line is being torqued to must be held in position with a second wrench to prevent twisting or damage to the secondary fitting or line.

Skills

1. **Make up a piece of rigid tubing that includes cutting it to the correct length, making a bend of the correct angle and radius, and correctly installing the type of fitting specified by the examiner.** (AM.I.D.S1)

2. **Make a proper single flare in a piece of aluminum alloy tubing.** (AM.I.D.S1)

3. **Make a proper double flare in a piece of aluminum alloy tubing.** (AM.I.D.S1)

4. **Make a proper bead in a piece of aluminum alloy tubing.** (AM.I.D.S1)

5. **Install a piece of rigid tubing in an aircraft, using the correct routing and approved mounting methods.** (AM.I.D.S2)

6. **Install and properly preset an MS flareless fitting on a piece of rigid tubing.** (AM.I.D.S2)

7. **Repair a piece of damaged rigid tubing by installing a union and the proper connecting fittings.** (AM.I.D.S2)

8. **Demonstrate the correct way to repair a damaged section of high-pressure rigid tubing using swaged fittings.** (AM.I.D.S2)

9. **Install a flexible hose in an aircraft, using the correct routing and approved mounting methods.** (AM.I.D.S3)

10. **Given examples of correct and incorrect flexible hose installation, identify the correct installations and explain to the examiner the problems with the incorrectly installed hoses.** (AM.I.D.S4)

11. **Given examples of acceptable and unacceptable bends in rigid tubing, identify the good bends and explain the reason the bad ones are not acceptable.** (AM.I.D.S5)

12. **Demonstrate the correct way to repair a damaged section of rigid tubing using AN flare fittings.** (AM.I.D.S5)

13. **Identify the lines and fittings used for rigid tubing.** (AM.I.D.S6)

14. **Identify the lines and fittings used for flexible hose.** (AM.I.D.S6)

15. **Fabricate a flexible hose using reusable fittings, pressure test it, and demonstrate its proper installation.** (AM.I.D.S7)

16. **Fabricate a flareless-fitting tube connection with the supplied component.** (AM.I.D.S8)

E. Aircraft Materials, Hardware, and Processes

References: AC 43.13-1; FAA-H-8083-30

Knowledge

1. **What is the alloy number of the most commonly used aluminum alloy for aircraft structural use?** (AM.I.E.K1)
   2024-T

2. **What is the alloy number of a high strength aluminum alloy that has zinc as an alloying component?** (AM.I.E.K1)
   7075

3. **What type of composite material is used when stiffness is the prime requirement?** (AM.I.E.K1)
   Graphite (carbon).

4. **What type of composite material is used when toughness is the prime requirement?** (AM.I.E.K1)
   Kevlar®.

5. **Why is a piece of steel tempered after it has been hardened?** (AM.I.E.K2)
   When steel is hardened, it becomes brittle. Tempering removes some of this brittleness.

6. **How is steel annealed?** (AM.I.E.K2)
   It is heated to just above its upper critical temperature until it reaches a uniform temperature throughout, then it is allowed to cool very slowly in the oven.

7. **How is steel hardened?** (AM.I.E.K2)
   It is heated to its critical temperature and quenched in water, brine, or oil.

8. **What is meant by tempering steel?** (AM.I.E.K2)
   The steel is first hardened; then some of the hardness is removed to relieve some of the internal stresses and brittleness.

9. **What is meant by case hardening?** (AM.I.E.K2)
   The surface of the metal is hardened by the infusion of carbon or aluminum nitride. The interior of the metal remains strong and tough.

10. **What are two methods of case hardening?** (AM.I.E.K2)
    Carburizing and nitriding.

11. **How is steel nitrided?** (AM.I.E.K2)
    The steel part is heated in a retort in which there is an atmosphere of ammonia (a compound of nitrogen and hydrogen). Aluminum, an alloying element in the steel, combines with the nitrogen to form an extremely hard aluminum nitride on the surface of the steel.

12. **What is the method of solution heat treatment of an aluminum alloy?** (AM.I.E.K2)
    The metal is hardened by heating it in a furnace to a specified temperature and immediately quenching it in water. It is soft when it is removed from the quench, but as it ages it regains its hardness.

13. **What is the method of precipitation heat treating of an aluminum alloy?** (AM.I.E.K2)
    The metal is heated and quenched, then it is returned to the oven and heated to a lower temperature. It is held at this temperature for a specified time, then removed from the oven and allowed to cool in still air. This increases the strength and hardness of the metal. Precipitation heat treating is also called artificial aging.

14. **What is another name for precipitation heat treatment?** (AM.I.E.K2)
    Artificial aging.

15. **Identify and explain the differences between heat-treated and non-heat-treated aluminum alloys.** (AM.I.E.K2)
    Heat-treated aluminum alloy designations are normally followed with a -T, followed by a number (1–10) that designates the type of heat treat. A common aircraft aluminum alloy is 2024-T3. Heat-treated aluminum is stronger than non-heat-treated aluminum.

16. **Why is it important that a piece of aluminum alloy be quenched immediately after it is removed from the heat treating oven?** (AM.I.E.K2)
    Any delay in quenching aluminum alloy after it is removed from the oven will allow the grain structure to grow enough that intergranular corrosion is likely to form in the metal.

17. **Describe and explain the forces placed on aircraft materials.** (AM.I.E.K3)
    a. *Tension*—A force tending to stretch or elongate the material.
    b. *Compression*—A force tending to compress or shrink.
    c. *Torsion*—A twisting force on an object.
    d. *Bending*—A perpendicular force on the longitudinal axis of an object, causing it to flex or bend. The inner surface of the bend will be in compression and the outer surface of the bend will be in tension.
    e. *Strain*—The stress on an object which can cause the object to change shape or to become distorted.
    f. *Shear*—A force that tries to cut or slice through an object. The force is parallel to the material cross section.

18. **What type of loading should be avoided when using a self-locking nut on an aircraft bolt?** (AM.I.E.K4)
    A self-locking nut should not be used for any application where there are any rotational forces applied to the nut or to the bolt.

19. **What determines the correct grip length of a bolt used in an aircraft structure?** (AM.I.E.K4)
    The grip length of the bolt should be the same as the combined thicknesses of the materials being held by the bolt.

20. **How tight should the nut be installed on a clevis bolt that is used to attach a cable fitting to a control surface horn?** (AM.I.E.K4)
    The nut on a clevis bolt should not be tight enough to prevent the clevis bolt from turning in the cable fitting and the horn.

21. **What bolt is described by this number: AN6-14A?** (AM.I.E.K4)
    AN6 = Hex head bolt, 6/16 (3/8) inch diameter.
    14 = Length = 1-4/8 (1-1/2) inch long.
    A = The shank is not drilled for a cotter pin.

22. **What is indicated by a triangle on the head of a steel bolt?** (AM.I.E.K4)
    This is a close tolerance bolt.

23. **What is a correct application for self-tapping sheet metal screws on an aircraft?** (AM.I.E.K4)
They may be used to hold nonstructural components onto the aircraft.

24. **How can you tell when a self-locking nut must be discarded?** (AM.I.E.K4)
When you can screw the nut onto a bolt without having to use a wrench.

25. **What is a channel nut?** (AM.I.E.K4)
A series of nuts mounted loosely in a channel that is riveted to the aircraft structure. You can install screws in a channel nut without having to hold the nut with a wrench.

26. **Of what two materials are cotter pins made?** (AM.I.E.K4)
Low-carbon steel and corrosion-resistant steel.

27. **What is the smallest size cable that is allowed to be used in the primary control system of an aircraft?** (AM.I.E.K4)
1/8-inch diameter.

28. **What type of control cable must be used when pulleys are used to change the direction of cable travel?** (AM.I.E.K4)
Extra-flexible cable (7 x 19).

29. **What is the purpose of safety wire?** (AM.I.E.K5)
To prevent a nut, bolt, or other component from loosening.

30. **After torquing a castellated nut, can the nut be loosened to align with the hole in a bolt for installing the cotter pin?** (AM.I.E.K5**)**
No. Never over-torque or loosen a torqued nut to align safety wire or cotter pin holes. The nut should be torqued to the minimum torque, and then tightened to align with the safety hole, without exceeding the maximum torque.

31. **What kind of measuring instrument is used to measure the runout of an aircraft engine crankshaft?** (AM.I.E.K6)
A dial indicator.

32. **What measuring instruments are used to measure the fit between a rocker arm shaft and its bushing?** (AM.I.E.K6)
The outside diameter of the shaft is measured with a micrometer caliper. The inside of the bushing is measured with a telescoping gauge and the same micrometer caliper.

33. **In what increments can a vernier micrometer caliper be read?** (AM.I.E.K6)
One ten thousandth (0.0001) inch.

34. **What is an advantage of a vernier caliper over a micrometer caliper?** (AM.I.E.K6)
The range of a vernier caliper is far greater than that of a micrometer caliper.

35. **What precision tool is used to measure piston ring side clearance?** (AM.I.E.K6)
Thickness gauge.

36. **What precision tools are used to measure the inside diameter of a cylinder?** (AM.I.E.K6)

    A telescoping gauge and a micrometer caliper.

37. **What precision tools are used to measure the inside diameter of a small hole?** (AM.I.E.K6)

    A small hole gauge and a micrometer caliper.

38. **What kind of solder is recommended for soldering electrical wires?** (AM.I.E.K7)

    60/40 resin-core solder.

39. **What is the function of the flux used in soldering?** (AM.I.E.K7)

    Flux covers the cleaned and heated metal to keep oxygen away from it so oxides cannot form. Oxides keep the solder from adhering to the surface of the metal.

40. **What are the three most common types of torque wrenches?** (AM.I.E.K8)

    Deflecting beam, dial indicating, and micrometer setting types.

41. **Where would an aviation mechanic find the appropriate hardware specifications for use on a particular aircraft?** (AM.I.E.K9)

    In the aircraft manufacturer's illustrated parts catalog (IPC).

42. **Where would an aviation mechanic find the material specifications for structural repairs on an aircraft?** (AM.I.E.K9)

    In the aircraft manufacturer's structural repair manual (SRM).

43. **Why is it very important that the surface of a piece of clad aluminum alloy not be scratched?** (AM.I.E.K9)

    The pure aluminum used for the cladding is noncorrosive, but the aluminum alloy below the cladding is susceptible to corrosion. If the cladding is scratched through, corrosion could form.

44. **What is the preload value when torquing a nut?** (AM.I.E.K10)

    Preload is the force or torque required to overcome the friction between the threads on the nut and the threads of the bolt prior to the nut contacting the mating surface.

45. **How is the preload value used when torquing a nut?** (AM.I.E.K10)

    The preload torque is added to the specified torque to obtain the required setting on the torque wrench.

46. **How can the material of a bolt be identified?** (AM.I.E.K11)

    By manufacturer markings that are on the bolt head.

47. **How is the welding flux removed from a piece of aluminum that has been gas-welded?** (AM.I.E.K12)

    It should be removed by scrubbing it with hot water and a bristle brush.

48. **What are the visual characteristics of a good weld?** (AM.I.E.K12)
A good weld is uniform in width; the ripples are even and well feathered into the base metal and show no burn due to overheating. The weld has good penetration and is free of gas pockets, porosity, or inclusions.

49. **What are the visual characteristics of a bad weld?** (AM.I.E.K13)
A bad weld has irregular edges and considerable variation in the depth of penetration. It often has the appearance of a cold weld.

50. **What must be done to a welded joint if it must be rewelded?** (AM.I.E.K14)
All traces of the old weld must be removed so the new weld will penetrate the base metal.

51. **What is meant by normalizing a piece of steel after it has been welded or machined?** (AM.I.E.K14)
Normalizing removes stresses that are locked into the material by welding or machining.

52. **How is a steel structure normalized after it has been welded?** (AM.I.E.K14)
Heat the steel structure to a temperature above its critical temperature and allow it to cool in still air.

53. **What determines the size of tip that is to be used when gas-welding steel?** (AM.I.E.K14)
The thickness of the material being welded. The size of the tip orifice determines the amount of flame produced and thus the amount of heat put into the metal.

Risk Management

1. **What are the common personal protection equipment (PPE) items that an aviation mechanic should have access to?** (AM.I.E.R1)
Hearing protection, eye protection, gloves, aprons, dust masks, and respirators.

2. **What are some of the dangers that could be caused by improper torque?** (AM.I.E.R2)
Too low of torque could cause components to loosen and come apart. Too high of torque could cause failure of the bolt or component or cause a rotating component to seize and not function as designed. All of these failures could cause loss of a system, or even loss of an aircraft and potential loss of life.

3. **Why is the use of used hardware or suspected unapproved parts (SUPS) undesirable?** (AM.I.E.R3)
In addition to creating an unairworthy condition, the risk of failure increases, which could cause damage to a system or the aircraft.

4. **Where would an aviation mechanic find the proper procedure for torquing critical, highly stressed fasteners?** (AM.I.E.R4)
In the aircraft manufacturer's maintenance manual.

5. **What is the proper technique when using a torque wrench?** (AM.I.E.R4)
   Pull the wrench assembly with a smooth, steady motion until the proper torque is reached. A fast or jerky motion will result in an improper torque.

Skills

1. **Properly safety wire two turnbuckles in an aircraft control system. Use the single-wrap method on one turnbuckle and the double wrap method for the other.** (AM.I.E.S1)

2. **Properly safety a series of bolts that are specified by the examiner.** (AM.I.E.S1)

3. **Properly install a cotter pin in a bolt fitted with a castellated nut.** (AM.I.E.S1)

4. **Determine the torque value for an aircraft fastener.** (AM.I.E.S2)

5. **Demonstrate the proper torque techniques.** (AM.I.E.S2)

6. **Inspect a set of parts with different types of welds. Identify any defective welds and explain the reason why the weld is not airworthy.** (AM.I.E.S3)

7. **Identify by the numbers stamped on the metal which aluminum alloys are heat treatable.** (AM.I.E.S4)

8. **What is meant by an icebox rivet?** (AM.I.E.S4)
   A rivet made of 2017 or 2024 aluminum alloy. These rivets are heat-treated and quenched, then stored in a sub-freezing ice box until they are ready to be used. The cold storage delays the hardening of the rivet.

9. **What are the two most commonly used rivet heads?** (AM.I.E.S4)
   MS20470 universal head and MS20426 100° countersunk head.

10. **What alloy is identified by the following rivet codes: A, AD, D, DD, B?** (AM.I.E.S4)
    A is 1100 aluminum.
    AD is 2117-T aluminum alloy.
    D is 2017-T aluminum alloy.
    DD is 2024-T aluminum alloy.
    B is 5056-T aluminum alloy.

11. **What mark on the head of a rivet identifies the following alloys: 2117-T, 2017-T, 2024-T, 5056-T?** (AM.I.E.S4)
    2117-T—recessed dot.
    2017-T—raised dot.
    2024-T—raised double dash.
    5056-T—raised cross.

12. **Given an assortment of threaded fasteners, identify a clevis bolt, close tolerance bolt, corrosion-resistant steel bolt, aluminum alloy bolt, machine screw, high-temperature self-locking nut, low-temperature self-locking nut, self-tapping sheet metal screw.** (AM.I.E.S5)

13. **Use a telescoping gauge and a micrometer caliper to measure the fit between a shaft and its bushing.** (AM.I.E.S6)

14. **Measure the diameter of a shaft to the nearest ten-thousandth of an inch, using a vernier micrometer caliper.** (AM.I.E.S6)

15. **Check the calibration of a micrometer.** (AM.I.E.S6)

16. **Locate the dimensional tolerances allowed for piston ring side clearance in an overhaul manual furnished by the examiner.** (AM.I.E.S6)

17. **Use a dial indicator to measure the runout of the crankshaft of an aircraft engine.** (AM.I.E.S7)

18. **Identify, by the number of strands of wire, a piece of extra-flexible control cable, and measure its diameter.** (AM.I.E.S8)

19. **Install a swaged fitting on a piece of aircraft control cable.** (AM.I.E.S9)

20. **Select the correct aluminum alloy for a specified aircraft structural repair.** (AM.I.E.S10)

21. **Explain to the examiner the reason aluminum alloy must not be exposed to excessively high temperature.** (AM.I.E.S10)

22. **Given an assortment of aircraft rivets, identify an AD rivet, a DD rivet, a D rivet, and an A rivet.** (AM.I.E.S11)

23. **Correctly install a Heli-Coil insert in an aluminum casting.** (AM.I.E.S11)

24. **Demonstrate to the examiner how to select the correct materials for a given repair, using the supplied aircraft manufacturer's manuals.** (AM.I.E.S12)

25. **Using the samples given, demonstrate to the examiner how to distinguish between heat-treated and non-heat-treated aluminum alloys.** (AM.I.E.S13)

26. **Demonstrate to the examiner how to check for proper calibration of a micrometer.** (AM.I.E.S14)

F. Ground Operations and Servicing

References: AC 43.13-1, AC 150/5210-20; FAA-H-8083-30

Knowledge

1. **What special precautions should be taken when towing an aircraft that is equipped with a steerable nose wheel?** (AM.I.F.K1)
   Be sure that the nose wheel does not try to turn past its stops. Some aircraft require the torsion links on the nose wheel strut to be disconnected when towing.

2. **What kind of rope is best to tie down an aircraft?** (AM.I.F.K2)
   Nylon or polypropylene.

3. **What kind of knot is used for securing an airplane with a rope?** (AM.I.F.K2)
   Bowline.

4. **What are the correct procedures to use when pressure fueling an aircraft?** (AM.I.F.K3)
   Most pressure fueling systems consist of a pressure fueling hose and a panel of controls and gauges that permit one person to fuel or defuel any or all fuel tanks of an aircraft. Each tank can be filled to a predetermined level. Always ensure that a ground is established between the fueling equipment and the aircraft prior to fueling.

   Prior to fueling, the person fueling must check the following:
   a. Ensure all aircraft electrical systems and electronic devices, including radar, are turned off.
   b. Do not carry anything in shirt pockets. These items could fall into the fuel tanks.
   c. Ensure no flame-producing devices are carried by anyone engaged in the fueling operation. A moment of carelessness could cause an accident.
   d. Ensure that the proper type and grade of fuel is used. Do not mix avgas and jet fuel.
   e. Ensure that all the sumps have been drained.
   f. Wear eye protection. Although generally not as critical as eye protection, other forms of protection, such as rubber gloves and aprons, can protect the skin from the effects of spilled or splashed fuel.
   g. Do not fuel aircraft if there is danger of other aircraft in the vicinity blowing dirt in the direction of the aircraft being fueled. Blown dirt, dust, or other contaminants can enter an open fuel tank, contaminating the entire contents of the tank.
   h. Do not fuel an aircraft when there is lightning within 5 miles.
   i. Do not fuel an aircraft within 500 feet of operating ground radar.

5. **What are the most common aviation gasoline colors and corresponding grades?** (AM.I.F.K3)
   Red—80; green—100; blue—100LL; purple—115.

6. **What publication lists the standard hand signals used for directing a taxiing aircraft?** (AM.I.F.K4)
   FAA *Aeronautical Information Manual* and *Aviation Maintenance Technician Handbook—General* (FAA-H-8083-30).

7. **What action should a mechanic take while taxiing an aircraft on a runway if the tower shines a flashing red light at him?** (AM.I.F.K4)

   Taxi the aircraft clear of the runway in use.

8. **When starting an aircraft engine equipped with a float carburetor, in what position should the carburetor heat control be placed?** (AM.I.F.K5)

   In the Cold position.

9. **What is meant by a liquid lock in the cylinder of an aircraft engine, and how is it cleared?** (AM.I.F.K5)

   Oil accumulates in the cylinders below the center line of the engine and prevents the piston from moving to the top of its stroke. To clear a liquid lock, remove one of the spark plugs and turn the engine crankshaft until all of the oil is forced out of the cylinder.

10. **What is the procedure if a hung start occurs when starting a turbojet engine?** (AM.I.F.K5)

    Terminate the starting operation and find the reason the engine would not accelerate as it should.

11. **How far ahead of an idling turbojet engine does the danger area extend?** (AM.I.F.K5)

    25 feet.

12. **What are the control tower light signals that may be used when taxiing or towing an aircraft, and what are their meanings?** (AM.I.F.K5)

    a. Flashing green—Cleared to taxi.
    b. Steady red—Stop.
    c. Flashing red—Taxi clear of runway in use.
    d. Flashing white—Return to starting point.
    e. Alternating red and green—Exercise extreme caution.

13. **What is meant by detonation in an aircraft reciprocating engine?** (AM.I.F.K5)

    Detonation is an uncontrolled burning of the fuel in the cylinder of an engine. It is an explosion, rather than a smooth burning.

14. **What type of fire extinguisher is best suited for extinguishing an induction fire in a reciprocating engine?** (AM.I.F.K6)

    Carbon dioxide (CO_2).

15. **What is proper way to extinguish an induction system fire that occurs when starting a reciprocating engine?** (AM.I.F.K6)

    Keep the engine running, which will blow the fire out. If this does not work, use a CO_2 fire extinguisher directed into the induction air inlet.

16. **When servicing an aircraft with air, nitrogen, oil, and other fluids, where are the proper procedures and equipment requirements found?** (AM.I.F.K7)

    Consult the specific aircraft maintenance manual.

17. **What are the three types of oxygen that are commercially available?** (AM.I.F.K8)

    Aviator's breathing, industrial, and medical.

18. **Is medical oxygen a suitable alternative for aviator's breathing oxygen?** (AM.I.F.K8)
No. Medical oxygen contains moisture that can freeze in the cold temperatures found at higher altitudes.

19. **What grade of aviation gasoline is dyed blue?** (AM.I.F.K9)
Low-lead Grade-100.

20. **What type of fuel is Jet-A?** (AM.I.F.K9)
A fuel with a heavy kerosine base, a flash point of 110–150°F, a freezing point of –40°F, and a heat energy content of 18,600 Btu/pound.

21. **What are two functions of tetraethyl lead that is added to aviation gasoline?** (AM.I.F.K10)
It increases the critical pressure and temperature of the fuel, and it acts as a lubricant for the valves.

22. **Why is it important that turbine fuel not be mixed with aviation gasoline used in an aircraft reciprocating engine?** (AM.I.F.K11)
The turbine engine fuel will cause the engine to detonate severely.

23. **What is the danger of using fuel that vaporizes too readily?** (AM.I.F.K11)
Vapor lock can occur in the fuel lines. This will shut off the flow of fuel to the engine.

24. **How can an aviation mechanic reduce the risk of leaving tools in an aircraft?** (AM.I.F.K12)
By using shadow boards or foam cutouts for each tool in their toolbox. Missing tools are easily identified by a quick review of the toolbox.

25. **Why is it important that the aircraft and the fuel truck be grounded together before the aircraft is fueled?** (AM.I.F.K13)
Grounding prevents the buildup of static electricity that causes sparks that could ignite the fuel vapors.

26. **How should aircraft parts be protected?** (AM.I.F.K14)
When stored on a shelf, sensitive equipment should remain in original packaging or on padded shelves with measures taken to prevent parts from rolling or falling off storage shelves. Electronic equipment should remain in antistatic packaging. When moving aircraft parts to be installed on an aircraft, care should be taken to prevent the part from falling or becoming damaged until installation is complete.

27. **Where can information be found specifying what personal protective equipment (PPE) should be used when working with specific hazardous materials?** (AM.I.F.K15)
In the Safety Data Sheet for the hazardous material.

28. **Why is FOD (foreign object damage) dangerous?** (AM.I.F.K16)
FOD can lead to damage and failure of aircraft systems and engines.

Risk Management

1. **Explain the precautions that should be used when towing an aircraft.** (AM.I.F.R1)
   a. Before the aircraft to be towed is moved, a qualified person must be in the flight deck to operate the brakes if the towbar breaks or becomes unhooked.
   b. Use the proper towbar.
   c. Observe tow steering limits and disconnect steering linkages when appropriate.
   d. Tow at reasonable speeds.
   e. Use wing walkers to verify wingtip clearance from obstacles.
   f. Close external doors and airstairs.
   g. Install gear-down lock pins (when available).
   h. Follow airport control tower or ground control instructions and clearances.

2. **What precautions should be used when connecting external power equipment to an airplane?** (AM.I.F.R2)

   Consult the aircraft manufacturer's recommendations. In most cases: (1) turn aircraft power off, (2) turn external power equipment off and ensure proper voltage has been selected, (3) connect external power to the aircraft, (4) turn external power on, and (5) turn aircraft power on.

3. **What documentation is required for a certificated aircraft to legally operate on automotive gasoline?** (AM.I.F.R3)

   There must be a Supplemental Type Certificate for the aircraft to operate on automotive gasoline.

4. **What hazard is associated with fueling or defueling ungrounded aircraft?** (AM.I.F.R3)

   Aviation fuel vapors exist when fueling or defueling an aircraft. During the process of fueling or defueling an aircraft, static electricity builds up due to the flow of fuel inside a rubber fueling hose and, if not grounded properly, can cause a spark that ignites the fuel vapors.

5. **What should be done to a reciprocating engine fuel system if turbine fuel is inadvertently put into the tanks and the engine is run?** (AM.I.F.R4)

   The entire fuel system must be drained and flushed with gasoline, the engine given a compression check, and all of the cylinders given a borescope inspection. The oil must be drained and all of the filters and strainers checked. After the aircraft is properly fueled, the engine must be given a proper run-up and check.

6. **What damage is likely to occur if an aircraft reciprocating engine that is designed to use grade 100 fuel is operated with grade 80 fuel?** (AM.I.F.R4)

   Detonation will occur, leading to bent connecting rods, burned pistons, and cracked cylinder heads.

7. **What information must be located near the fuel tank filler opening in an aircraft powered by a reciprocating engine?** (AM.I.F.R4)

   The word "Avgas," the minimum fuel grade.

8. **What information must be located near the fuel tank filler opening in an aircraft powered by a turbine engine?** (AM.I.F.R4)

   The words "Jet Fuel," the permissible fuel designations, or references to the airplane flight manual (AFM) for permissible fuel designations; additionally, for pressure fueling systems, the maximum permissible fueling supply pressure and the maximum permissible defueling pressure.

9. **What precaution must be observed regarding the tools used when servicing an aircraft oxygen system?** (AM.I.F.R5)

   Be sure that there is no oil or grease on the tools. Pure oxygen can cause spontaneous combustion of petroleum products. Two personnel should be used; one for controlling the valves of the servicing equipment and the other to monitor the pressure in the aircraft system. Do not service the oxygen system when fueling, defueling, or conducting any maintenance that could provide a source of ignition.

10. **Why is it important to use a checklist when starting and running an aircraft engine?** (AM.I.F.R6)

    Using a checklist will lead to safe conditions, protecting the aircraft, surrounding equipment, and personnel.

11. **Where are the procedures found to start an aircraft engine correctly and safely?** (AM.I.F.R7)

    In the specific aircraft manufacturer's aircraft maintenance manual or aircraft flight manual.

12. **Explain how to mitigate risk when starting and operating an engine while troubleshooting or adjusting engine controls.** (AM.I.F.R8)

    Always follow the specific aircraft maintenance manual instructions and cautions. Adjustments to the engine should be made only when the engine is not running.

13. **What are the dangers of operating an aircraft engine with cowling removed contrary to manufacturer instructions?** (AM.I.F.R9)

    The aircraft cowling may become damaged, the engine may not receive enough cooling air and overheat, and/or damaged cowling may come loose and damage the engine, propeller, or aircraft.

14. **What are some of the risks associated with operating an aircraft in the vicinity of other aircraft or ground support equipment, and how can the risks be mitigated?** (AM.I.F.R10)

    Always aim an aircraft so that the blast from the propellers or engines will not damage other aircraft, equipment, or personnel.

Skills

1. **Demonstrate to the examiner how to perform a foreign object damage control procedure for the area or task given.** (AM.I.F.S1)

2. **Properly connect auxiliary power to an aircraft for the purpose of starting the engine. Explain to the examiner the safety precautions that should be taken.** (AM.I.F.S2)

3. **Connect a tow bar to an aircraft nose wheel for moving the aircraft. Explain to the examiner the precautions that must be taken when moving the aircraft.** (AM.I.F.S3)

4. **Demonstrate to the examiner the correct hand signals to use when directing the operator of an aircraft to:**
   a. **Start engine number one.**
   b. **Move the aircraft ahead.**
   c. **Stop the aircraft.**
   d. **Emergency stop the aircraft.**
   e. **Shut the engine down.**

   (AM.I.F.S4)

5. **Drain a sample of fuel from an aircraft fuel system. Check the fuel for the presence of water and identify the grade of the fuel.** (AM.I.F.S5)

6. **Drain a sample of fuel from one of the wing tanks and from the main fuel strainer. Explain to the examiner the way to check the fuel for contaminants.** (AM.I.F.S5)

7. **Identify the different grades of aviation gasoline.** (AM.I.F.S6)

8. **Fuel an aircraft using the proper grade and amount of fuel. Explain to the examiner the safety procedures that must be used for this operation.** (AM.I.F.S7, AM.I.F.S8)

9. **Determine the amount of fuel remaining in an aircraft.** (AM.I.F.S8)

10. **Properly start, run up, and shut down an aircraft reciprocating engine.** (AM.I.F.S9)

11. **Properly start, run up, and shut down an aircraft turbine engine.** (AM.I.F.S9)

12. **Demonstrate to the examiner the correct way to clear the cylinders of an aircraft reciprocating engine of a hydraulic lock.** (AM.I.F.S9)

13. **Check the oil supply and determine for the examiner the correct grade of oil to use.** (AM.I.F.S9)

14. **Taxi an aircraft, using the proper safety procedures.** (AM.I.F.S9)

15. **List the procedures for extinguishing an engine induction fire during starting.** (AM.I.F.S10)

16. **Demonstrate the correct method of securing an aircraft to the flight line tie-downs.** (AM.I.F.S11)

17. **Properly secure an airplane for overnight storage on an outside tie-down area.** (AM.I.F.S11)

18. **Properly secure a helicopter for overnight storage in an outside parking area accounting for high wind conditions.** (AM.I.F.S11)

19. **Demonstrate to the examiner the proper way to tie an aircraft down.** (AM.I.F.S11)

20. **Demonstrate to the examiner the proper way to prepare a tied-down aircraft or helicopter for predicted high winds.** (AM.I.F.S11)

21. **Secure a turbine-powered aircraft after engine shutdown.** (AM.I.F.S12)

G. Cleaning and Corrosion Control

References: AC 43.13-1; FAA-H-8083-30

Knowledge

1. **What is used to clean transparent plastic windshields and windows of an aircraft?** (AM.I.G.K1)
   Mild soap and lots of clean water.

2. **What is used to neutralize the electrolyte from a lead-acid battery that has been spilled on an aircraft structure?** (AM.I.G.K1)
   A solution of bicarbonate of soda and water.

3. **What is used to neutralize the electrolyte from a nickel-cadmium battery that has been spilled on an aircraft structure?** (AM.I.G.K1)
   A solution of boric acid and water, or vinegar.

4. **What solvent is recommended for removing grease from aircraft fabric prior to doping it?** (AM.I.G.K1)
   Methyl-ethyl-ketone (MEK) or lacquer thinner.

5. **What is the proper way to clean plastics?** (AM.I.G.K1)
   Flush the plastic with clean water. Remove dirt by hand. Plastics are easily scratched and usage of rags is not recommended.

6. **What causes corrosion?** (AM.I.G.K2)
   Corrosion is a natural phenomenon that attacks metal by chemical or electrochemical action and converts it into a metallic compound, such as an oxide, hydroxide, or sulfate. Corrosion occurs because of the tendency for metals to return to their natural state. Noble metals like gold and platinum do not corrode since they are chemically uncombined in their natural state. Conditions that can cause corrosion to develop or worsen are moisture (acts as an electrolyte), oxygen (acts as a catalyst), chemicals (initiate or catalyze the electromechanical process), and temperature (accelerates the process).

7. **Where is filiform corrosion most likely to occur on an aircraft?** (AM.I.G.K3)
   Under a dense coating of topcoat enamel such as polyurethane. Filiform corrosion is caused by improperly cured primer.

8. **Where is fretting corrosion most likely to occur on an aircraft?** (AM.I.G.K3)
   In a location where there is a slight amount of relative movement between two components and no way for the corrosive residue to be removed as it forms.

9. **Where is intergranular corrosion most likely to occur on an aircraft?** (AM.I.G.K3)
   Along the grain boundaries of aluminum alloys that have been improperly heat-treated. Extruded aluminum alloy is susceptible to intergranular corrosion.

10. **Where is stress corrosion most likely to occur on an aircraft?** (AM.I.G.K3)
    In any metal component that is continually under a tensile stress. The metal around holes in castings fitted with pressed-in bushings is susceptible to stress corrosion.

11. **What areas of an aircraft are most prone to corrosion?** (AM.I.G.K4)
    Battery compartment, exhaust system and exhaust trails, wheel wells, lower area of the belly (bilge), piano hinges, areas of dissimilar metal contact, welded areas, inside of fuel tanks (especially integral tanks), metal fittings under high stress, lavatories, and food service areas.

12. **What is used to keep corrosion from forming on structural aluminum alloy?** (AM.I.G.K5)
    An oxide coating or aluminum cladding.

13. **How should an A&P minimize corrosion at piano hinges?** (AM.I.G.K5)
    Piano hinges should be kept as clean and dry as practicable and lubricated with a low-viscosity moisture dispersing agent.

14. **How is the inside of structural steel tubing protected from corrosion?** (AM.I.G.K5)
    The tubing is filled with hot linseed oil and then drained.

15. **How does filiform corrosion usually appear on an aircraft structure?** (AM.I.G.K6)
    As thread-like lines of puffiness under a film of polyurethane or other dense finish system topcoats.

16. **What must be done to a piece of aluminum alloy to remove surface corrosion and to treat the metal to prevent further corrosion?** (AM.I.G.K7)
    Remove the corrosion residue with a bristle brush or a nylon scrubber. Neutralize the surface with chromic acid or with some type of conversion coating. Protect the surface from further corrosion with a coat of paint.

17. **How may rust be removed from a highly stressed metal part?** (AM.I.G.K7)
    By glass bead blasting, by careful polishing with mild abrasive paper, or by using fine buffing compound on a cloth buffing wheel.

18. **What tools are proper for removing corrosion from aluminum alloy?** (AM.I.G.K7)
    Aluminum wool or aluminum wire brushes. Severe corrosion can be removed with a rotary file.

19. **What type of device is used to remove surface corrosion from a piece of magnesium alloy?** (AM.I.G.K7)
    A stiff hog-bristle brush.

20. What must be done to a piece of aluminum alloy to remove surface corrosion and to treat the metal to prevent further corrosion? (AM.I.G.K7)

Remove the corrosion residue with a bristle brush or a nylon scrubber. Neutralize the surface with chromic acid or with some type of conversion coating. Protect the surface from further corrosion with a coat of paint.

21. How is rust removed from a highly stressed metal part? (AM.I.G.K7)

By glass bead blasting, by careful polishing with mild abrasive paper, or by using fine buffing compound on a cloth buffing wheel.

22. What is a corrosion preventive compound (CPC)? (AM.I.G.K8)

CPC is a waxy sealant or thin-film dielectric that can be sprayed into aircraft structures to coat the surfaces and work its way between skins to prevent corrosion.

23. What type of environment would CPCs normally be used in? (AM.I.G.K9)

CPCs are used in corrosive environments such as aerial agricultural spray operations and saltwater environments.

24. How are CPCs applied to aircraft structures? (AM.I.G.K10)

CPCs are normally applied with high-pressure spray equipment that creates a fog of the CPC material inside the aircraft structure. Long spray wands are used to reach difficult-to-access areas.

25. What is used to clean a composite material prior to repair? (AM.I.G.K11)

Acetone or methyl-ethyl-ketone (MEK) is commonly used. It is important to verify that the solvent residue is compatible with the resin system.

26. Where is dissimilar metal corrosion most likely to occur on an aircraft? (AM.I.G.K12)

Anywhere different types of metal come in contact with each other, especially where moisture is present.

27. What can be used to repair the anodized surface of an aluminum alloy part? (AM.I.G.K13)

A chemical conversion coating such as Alodine.

28. What are three types of primer that may be used when painting an aircraft? (AM.I.G.K14)

Zinc chromate primer, wash primer, and epoxy primer.

29. What type of thinner is used with zinc chromate primer? (AM.I.G.K14)

Toluol or toluene.

30. What type of primer is used when the maximum protection of the metal is required? (AM.I.G.K15)

Epoxy primer.

31. What are some different types of primer materials? (AM.I.G.K15)

Wash primers, red iron oxide, gray enamel undercoat, urethane, epoxy, zinc chromate.

32. **How thick should a coat of wash primer be that is used on an aluminum alloy aircraft structure?** (AM.I.G.K15)
It should be thin enough that it does not hide the surface of the metal.

33. **What are some different types of topcoat materials?** (AM.I.G.K16)
Dope, synthetic enamel, lacquers, polyurethane, urethane, acrylic urethanes.

34. **What is the primary purpose of a finish on a composite material?** (AM.I.G.K16)
To protect the composite structure from ultraviolet degradation.

35. **What are the methods used to remove finishes from a composite material?** (AM.I.G.K17)
By light sanding or abrasive blast with plastic media.

36. **When mixing epoxy paint, should the converter be added to the resin or the resin to the converter?** (AM.I.G.K17)
The converter should always be added to the resin, never the resin to the converter.

37. **What should be done to a corroded aluminum alloy surface after the corrosion has been removed by mechanical methods?** (AM.I.G.K17)
The surface should be treated with a chemical conversion coating before the topcoats are applied.

38. **What may be done to the surface to prevent filiform corrosion beneath the topcoat?** (AM.I.G.K17)
Be sure that the primer is properly and completely cured.

39. **How is the finish removed from a fiberglass aircraft component that is being repaired?** (AM.I.G.K17)
The finish must be sanded off. Paint remover can soften the resin of which the component is made.

40. **How can a vinyl film decal be removed from an aluminum alloy surface?** (AM.I.G.K17)
Place a cloth saturated with cyclohexanone or MEK over the decal until it is softened, and scrape it off of the surface with a plastic scraper.

41. **Why is retarder used in dope when the dope is being sprayed in humid conditions?** (AM.I.G.K18)
The retarder slows the drying of the dope and keeps it from blushing.

42. **What is meant by dope blushing, and what causes it?** (AM.I.G.K18)
Blushing is a condition in a dope finished component in which moisture from the atmosphere condenses on the surface and causes some of the cellulose to precipitate from the finish. Blushing leaves a porous, dull, and weak finish.
Blushing may be caused by the temperature being too low, the humidity being too high, or by drafts or sudden changes in the temperature.

43. **What causes blushing in a dope finish?** (AM.I.G.K18)
Too high humidity.

44. **What causes fisheyes in the finish?** (AM.I.G.K19)
Localized surface contamination.

45. **Where are the registration marks required to be placed on a fixed-wing aircraft?** (AM.I.G.K20)
On a vertical tail surface or on the side of the fuselage.

46. **What is the generally required dimensions of the registration numbers on the side of a fixed-wing aircraft?** (AM.I.G.K20)
12-inches tall and 2/3 as wide as they are high. The letters M and W may be as wide as they are high. The numeral 1 is 1/6 as wide as it is high.

47. **What is the regulation regarding the color of the registration marks?** (AM.I.G.K20)
The color must contrast with the background and be legible.

48. **What is a cause of poor adhesion between the topcoat and the fill coats on a fabric surface?** (AM.I.G.K21)
Too much aluminum powder in the aluminum pigmented dope.

49. **What is a cause of a rough finish on a freshly sprayed surface?** (AM.I.G.K21)
Too high atomizing air pressure on the spray gun.

50. **What causes sags and runs on a surface that has just been sprayed?** (AM.I.G.K21)
Too much finishing material being applied in one coat.

51. **What causes orange peel, or spray mottle, in a finish?** (AM.I.G.K21)
Incorrect paint viscosity or improper setting of the spray gun.

52. **What causes pinholes in a finish?** (AM.I.G.K21)
Excessive atomizing air on the spray gun.

53. **What safety precaution must be observed when sweeping a paint room that has dried dope or lacquer overspray on the floor?** (AM.I.G.K22)
The floor must be wet down with water before it is swept. Static electricity from dry sweeping can cause a fire.

54. **What personal protection equipment (PPE) should be used while painting?** (AM.I.G.K22)
Respirator, eye protection, and skin protection.

55. **Why are some portions of the structure of an aircraft dope proofed before they are covered with fabric?** (AM.I.G.K23)
Dope proofing keeps the fabric from sticking to the structure when the first coat of dope is applied. The fabric normally sags enough to touch the structure before it begins to pull taut.

56. **What are some key application techniques for a quality paint job?** (AM.I.G.K23)
Correct spray pressures and spray gun setting, proper spray gun handling (even, overlapping spray patterns, spray gun distance), and good lighting.

57. **Why should control surfaces be balanced after painting?** (AM.I.G.K24)
    The weight or distribution of the paint may have changed, and the control surface may be out of tolerance.

Risk Management

1. **Explain the health concerns that need to be addressed when using paints, solvents, and finishing materials.** (AM.I.G.R1)
   Hazardous materials may cause short-term or long-term damage to an individual. Always use appropriate PPE.

2. **Describe the risks associated with poor ventilation when painting.** (AM.I.G.R2)
   Poor ventilation will cause higher concentrations of vapors, some of which can be very volatile. High concentrations of paint in the air will also lead to overspray on the aircraft, leading to a poor-quality paint job.

3. **What risks or hazards are present if materials or processes for corrosion cleaning and treatment are not followed correctly?** (AM.I.G.R3)
   - Some chemicals used for cleaning metal or stripping paint can cause damage to undamaged paint, systems, or aircraft windows.
   - The application of a topcoat over an area where corrosion has not fully been removed may trap corrosion or moisture, leading to further corrosion.
   - The incorrect method of removing corrosion may cause severe damage to a part or structural assembly.

4. **Why is it important to have access to the SDS (Safety Data Sheet) for the products used during removal and treatment of corrosion?** (AM.I.G.R4)
   The SDS provides specific guidance on what PPE should be used (such as gloves, eye protection, respirators, etc.) when working through the corrosion removal and treatment process.

5. **Explain the precautions that should be taken when working with flammable chemicals.** (AM.I.G.R5)
   Always follow the guidance found in the SDS. Use proper containers, ventilation, and PPE, and have the correctly rated fire extinguishers on hand. Store or dispose of cleaning rags in appropriately rated fireproof containers.

6. **Describe the proper disposal of chemicals and waste materials.** (AM.I.G.R6)
   Chemicals and waste materials must be disposed of in accordance with Environmental Protection Agency (EPA) regulations and local laws and ordinances.

7. **What are the dangers of not using the proper PPE when working with paints and solvents?** (AM.I.G.R7)
   Paints and solvents can cause both short-term and long-term damage to an individual's health, up to, and including death.

8. **Explain how to mitigate the risks and hazards associated with the application of finishing materials.** (AM.I.G.R8)

   Always use an appropriately rated painting booth that filters and contains vapors and solvents. Use the correct PPE, which often includes a full painting suit with gloves, eye protection, and a respirator.

Skills

1. **Given samples of corroded aircraft structural materials, identify the type of corrosion. Describe the correct procedure for removing the corrosion and treating the damaged area to prevent further corrosion.** (AM.I.G.S1)

2. **Select the proper cleaning materials and remove grease and exhaust deposits from an aircraft structure.** (AM.I.G.S2)

3. **Select the proper cleaning materials and remove oil that has been spilled on an aircraft tire.** (AM.I.G.S2)

4. **Clean aluminum or magnesium parts with a caustic cleaner.** (AM.I.G.S2)

5. **Clean an assigned area of an aircraft.** (AM.I.G.S2)

6. **Apply a protective coating to a metallic material.** (AM.I.G.S3)

7. **Treat the cylinders of a reciprocating engine to prevent rust and corrosion when the engine is being prepared for long-time storage.** (AM.I.G.S3)

8. **Treat a piece of welded steel tubular structure to prevent rust and corrosion inside the tubing.** (AM.I.G.S3)

9. **Treat a piece of aircraft structure so moisture cannot reach the metal and cause corrosion.** (AM.I.G.S3)

10. **Locate the procedures for preparing a specified aircraft or aircraft parts for extended storage.** (AM.I.G.S3)

11. **Remove corrosion from a lead-acid battery box and treat the box to prevent further corrosion.** (AM.I.G.S3)

12. **Remove the corrosion from a piece of aluminum alloy furnished by the examiner and treat the metal to prevent further corrosion.** (AM.I.G.S3)

13. **Demonstrate to the examiner the correct way to remove rust from a highly stressed engine component.** (AM.I.G.S3)

14. **Inspect an acrylic nitrocellulose lacquer finish and describe your findings to the examiner.** (AM.I.G.S4)

15. **Inspect a painted surface for defects. Identify the type(s) of defects and explain to the examiner the proper corrective action.** (AM.I.G.S4)

16. **Inspect the supplied aircraft section or compartment for corrosion.** (AM.I.G.S5)

17. **Properly remove the finish from a piece of painted fiberglass-reinforced plastic material.** (AM.I.G.S6)

18. **Remove scratches from a plastic windshield or window.** (AM.I.G.S6)

19. **Explain to the examiner the correct size and location for the identification numbers that are required on an aircraft.** (AM.I.G.S7)

20. **Demonstrate the way to lay out the registration marks that are specified by the examiner.** (AM.I.G.S7)

21. **Properly remove the finish from a piece of fiberglass-reinforced aircraft structure so the structure can be repaired.** (AM.I.G.S8)

22. **Properly prepare a composite or metallic structure for painting and identify all treatment and coating materials required.** (AM.I.G.S8)

23. **Select the correct type of surface treatments for aluminum and steel surfaces.** (AM.I.G.S9)

24. **Identify multiple types of aircraft finishes.** (AM.I.G.S9)

25. **Demonstrate to the examiner the correct way to spray a surface with a polyurethane enamel.** (AM.I.G.S10)

26. **Properly adjust the pressure of the air on a spray gun and pressure pot for spraying aircraft dope.** (AM.I.G.S10)

27. **Properly remove the finish from a piece of painted aluminum alloy.** (AM.I.G.S11)

28. **Mechanically remove paint from a corroded aircraft part and determine the extent of the corrosion.** (AM.I.G.S11)

29. **Properly clean a piece of aluminum alloy and apply the correct amount of primer for best adherence of the topcoat material.** (AM.I.G.S11)

30. **Properly mix and apply a paint system specified by the examiner.** (AM.I.G.S12)

31. **Apply paint to a fiber-reinforced plastic material.** (AM.I.G.S12)

32. **Identify the correct thinner to use with a list of finishing materials furnished by the examiner.** (AM.I.G.S12)

33. **Determine if a refinished control surface requires balancing.** (AM.I.G.S12)

H. Mathematics

References: AC 43.13-1; FAA-H-8083-30

Knowledge

1. **What formula is used to find the area of a circle?** (AM.I.H.K1)
   $A = \frac{\pi}{4} \times D^2$ or, $A = \pi \times R^2$

2. **What formula is used to find the area of a rectangle?** (AM.I.H.K1)
   $A = L \times W$

3. **What formula is used to find the area of a triangle?** (AM.I.H.K1)
   $A = (B \times H) \div 2$

4. **What formula is used to find the volume of a cylinder?** (AM.I.H.K2)
   $V = \frac{\pi}{4} \times D^2 \times H$

5. **What formula is used to find the volume of a rectangular solid?** (AM.I.H.K2)
   $V = L \times W \times H$

6. **What is meant by the root of a number?** (AM.I.H.K3)
   The root of a number is one of two or more equal numbers that, when multiplied together, will produce the number.

7. **What is an angle?** (AM.I.H.K3)
   A figure formed by two lines radiating from a common point.

8. **What is meant by a right angle?** (AM.I.H.K3)
   An angle of 90°.

9. **What is an obtuse angle?** (AM.I.H.K3)
   An angle greater than 90° but less than a straight line.

10. **What is an acute angle?** (AM.I.H.K3)
    An angle between 0° and 90°.

11. **What is a quadrant?** (AM.I.H.K3)
    One fourth of a circle.

12. **What is a sector?** (AM.I.H.K3)
    The portion of a circle bounded by two radii and one of the intercepted arcs.

13. **What is a polygon?** (AM.I.H.K3)
    A closed figure having three or more sides.

14. **What is an acute triangle?** (AM.I.H.K3)
    A triangle with three acute angles.

15. **What is an obtuse triangle?** (AM.I.H.K3)
A triangle with one obtuse angle.

16. **What is a right triangle?** (AM.I.H.K3)
A triangle containing a 90° angle.

17. **What is an equilateral triangle?** (AM.I.H.K3)
A triangle with three equal sides.

18. **What is an isosceles triangle?** (AM.I.H.K3)
A triangle with two equal sides.

19. **What is a scalene triangle?** (AM.I.H.K3)
A triangle in which no two sides are equal.

20. **What is the hypotenuse of a right triangle?** (AM.I.H.K3)
The side opposite the right angle.

21. **What is a quadrilateral?** (AM.I.H.K3)
A polygon having four sides.

22. **What is a parallelogram?** (AM.I.H.K3)
A closed four-sided figure in which the opposite sides are parallel.

23. **What is a trapezoid?** (AM.I.H.K3)
A closed four-sided figure in which only one pair of opposite sides are parallel.

24. **What is a hexagon?** (AM.I.H.K3)
A closed figure with six equal sides.

25. **What is the significance of the constant π?** (AM.I.H.K3)
Pi (π) is the constant amount the circumference of a circle is greater than its diameter.

26. **What is the value of the constant π?** (AM.I.H.K3)
$\pi = 3.1416$

27. **What is meant by a ratio and how is a ratio expressed?** (AM.I.H.K4)
A fraction that compares one number to another; for example, 6:2 is a ratio.

28. **What is meant by a proportion?** (AM.I.H.K5)
A statement of equality between two ratios; for example, $\frac{a}{b} = \frac{b}{c}$ is a proportion.

29. **What is meant by a percentage?** (AM.I.H.K5)
A fraction having 100 as the denominator. For example: $65\% = \frac{65}{100}$

30. **What is the square root of 4,096?** (AM.I.H.K6)
64

31. **What is the square of 99?** (AM.I.H.K6)
9,801

32. **What is the cube of 5?** (AM.I.H.K6)
125

33. **What is the cube root of 1,000?** (AM.I.H.K6)
10

34. **What is 50 gallons converted to liters?** (AM.I.H.K7)
189.3 liters.

35. **What is 25 kilos converted to pounds?** (AM.I.H.K7)
55 pounds.

36. **What is the base in the following expression: 10^3?** (AM.I.H.K8)
10

37. **What is the exponent in the following expression: 10^3?** (AM.I.H.K8)
3

38. **What is meant by scientific notation?** (AM.I.H.K8)
A method of writing very large or very small numbers using powers of 10.

39. **How do you write 56,000,000 in scientific notation?** (AM.I.H.K8)
When converting standard notation to scientific notation, move the decimal to the left or right until there is one non-zero integer on its left. If you moved the decimal to the left, you'll multiply by ten times the number you moved it to the left, so 56,000,000 would be 5.6×10^7.

40. **How do you write 0.000 000 96 in scientific notation?** (AM.I.H.K8)
9.6×10^{-7}

41. **What is 57,435 rounded to the nearest 100?** (AM.I.H.K9)
57,400

42. **What is 3.141592 rounded to the nearest thousandth decimal point?** (AM.I.H.K9)
3.142

43. **What is the eighth power of 2?** (AM.I.H.K10)
256

44. **What is the official system of measurement used throughout the world, especially in scientific notation?** (AM.I.H.K11)
The metric system.

45. **What is meant by a negative number?** (AM.I.H.K12)
A number less than 0, or a number preceded by a minus (–) sign.

46. **What steps are used to add signed numbers?** (AM.I.H.K12)
    1. Add all numbers having a plus sign.
    2. Add all numbers having a minus sign.
    3. Of these answers, subtract the smaller from the larger number. The sign of the answer will be the sign of the larger number.

47. What steps are used to subtract signed numbers? (AM.I.H.K12)

1. Change the sign of the subtrahend (the number that is being subtracted).
2. If the signs are alike, add the two numbers and attach the common sign.
3. If the signs are different, subtract the smaller from the larger and give the answer the sign of the larger.

48. What steps are used to multiply signed numbers? (AM.I.H.K12)

1. Multiply the numbers, disregarding the signs.
2. If the signs of the two numbers are alike, the sign of the answer is positive.
3. If the signs of the two numbers are not alike, the sign of the answer is negative.

49. What steps are used to divide signed numbers? (AM.I.H.K12)

1. Disregard the signs and divide as though the numbers had no signs.
2. If the signs of the two numbers are alike, the sign of the answer is positive.
3. If the signs of the two numbers are not alike, the sign of the answer is negative.

50. What is the precedence of operations when solving an algebraic equation? (AM.I.H.K13)

1. Grouped expressions, such as parenthesis and radical expressions (square roots)
2. Exponents
3. Multiplication and division
4. Addition and subtraction

51. What is the sum of (+6) + (+12)? (AM.I.H.K13)

18

52. What is the sum of (+6) + (–12)? (AM.I.H.K13)

–6

53. What is the difference between (+6) – (+12)? (AM.I.H.K13)

–6

54. What is the difference between (–6) – (+12)? (AM.I.H.K13)

–18

55. What is the difference between (–6) – (–12)? (AM.I.H.K13)

6

56. What is the product of (–6) × (–12)? (AM.I.H.K13)

72

57. What is the product of (–6) × (+12)? (AM.I.H.K13)

–72

58. What is the quotient of (–6) ÷ (+12)? (AM.I.H.K13)

–½

59. What is the quotient of (–6) ÷ (–12)? (AM.I.H.K13)

½

Risk Management

1. **What are the risks associated with incorrectly applying the precedence of operations when solving algebraic equations?** (AM.I.H.R1)

   Incorrectly solving algebraic equations can have serious consequences. One example is incorrectly computing the center of gravity (CG) after weighing an aircraft. An incorrectly computed CG could lead to a catastrophic outcome if the aircraft is loaded outside of the actual allowable limits.

2. **What are the dangers of incorrectly applying the rules of adding and subtracting positive and negative integers?** (AM.I.H.R2)

   When installing and removing equipment from an airplane, it is important that the calculations are correct. Components may be either ahead or behind the datum and thus the moments can be either positive or negative. Incorrectly applying the rules of positive and negative integers will cause the stated weight and balance of the airplane to be incorrect.

3. **What are the dangers of incorrectly rounding off calculations?** (AM.I.H.R3)

   Rounding off calculations too far or incorrectly can lead to many dangerous situations. This can result in replacement wires that are too small, switches and circuit breakers that are not rated sufficiently for the load of the circuit, incorrectly sized alternators, and many other dangerous situations.

Skills

1. **Find the square and cube of a list of numbers.** (AM.I.H.S1)

2. **Find the square root of a list of numbers.** (AM.I.H.S1)

3. **Find the cube root of a list of numbers.** (AM.I.H.S1)

4. **Using the correct formula, find the volume of a cylinder of an aircraft engine with the bore and stroke specified by the examiner.** (AM.I.H.S2)

5. **Using the proper formula, find the area of a rectangle with the dimensions given by the examiner.** (AM.I.H.S3)

6. **Using the proper formula, find the area of a square with the dimensions given by the examiner.** (AM.I.H.S3)

7. **Using the proper formula, find the area of a triangle with the dimensions given by the examiner.** (AM.I.H.S3)

8. **Using the proper formula, find the area of a circle with the dimensions given by the examiner.** (AM.I.H.S3)

9. **Using the proper formula, find the area of a trapezoid with the dimensions given by the examiner.** (AM.I.H.S3)

10. **Find the area of a rectangular airfoil with the span and chord specified by the examiner.** (AM.I.H.S3)

11. **Using the correct formula, find the volume of a sphere with the diameter specified by the examiner.** (AM.I.H.S4)

12. **Using the correct formula, find the volume of a cube with the length of the side specified by the examiner.** (AM.I.H.S4)

13. **Find the volume of a baggage compartment with the dimensions given by the examiner.** (AM.I.H.S4)

14. **Convert a list of common fractions into percentages.** (AM.I.H.S5)

15. **Convert a number of percentages into common fraction with the smallest denominator.** (AM.I.H.S5)

16. **Using two numbers given by the examiner, determine the ratio between the two numbers.** (AM.I.H.S6)

17. **Find the speed of rotation of the shaft of a driven gear when the gear ratio and the rotational speed of the drive gear are given.** (AM.I.H.S6)

18. **Compute the change in torque value when using a torque wrench extension.** (AM.I.H.S6)

19. **Find the number that is a given percentage of another number.** (AM.I.H.S6)

20. **Find the cost of an object when the selling price and the profit percentage are known.** (AM.I.H.S6)

21. **Find the number of ounces of material A to be used with 6 ounces of material B, if the materials are to be mixed so the proportion is 6:1 (6A to 1B).** (AM.I.H.S6)

22. **Using the correct formula, determine the compression ratio with the dimensions given by the examiner.** (AM.I.H.S7)

23. **Using the correct formula, determine the torque values when converting from inch-pounds to foot-pounds and from foot-pounds to inch-pounds with the values provided by the examiner.** (AM.I.H.S8)

I. Regulations, Maintenance Forms, Records, and Publications

References: 14 CFR; AC 43.13-1, AC 43-9; FAA-H-8083-30

Knowledge

1. **Who is authorized to perform a 100-hour inspection on an aircraft?** (AM.I.I.K1)
   A certificated Aviation Mechanic who holds an Airframe and Powerplant rating.

2. **Who is authorized to perform an annual inspection on an aircraft?** (AM.I.I.K1)
   A certificated A&P mechanic who holds an Inspection Authorization.

3. **Can a certificated A&P mechanic supervise an unlicensed person as the unlicensed person performs a 100-hour inspection on an aircraft?** (AM.I.I.K1)
   No, a certificated mechanic must personally perform the inspection.

4. **Who is authorized to rebuild an aircraft engine and issue a zero time maintenance record?** (AM.I.I.K1)
   Only the manufacturer of the engine or a facility approved by the manufacturer.

5. **Which FAR specifies required maintenance records?** (AM.I.I.K1)
   14 CFR §91.417 specifies you must keep records of maintenance, preventive maintenance, major repairs and alterations, and 100-hour, annual, and progressive inspections.

6. **Are instructions included in the Federal Regulations mandatory or optional?** (AM.I.I.K1)
   They are mandatory.

7. **Who is authorized to approve an aircraft for return to service after a major repair?** (AM.I.I.K1)
   An aircraft mechanic holding an Inspection Authorization.

8. **Who is authorized to perform preventive maintenance on an aircraft that is not flown under Part 121, 127, 129, or 135?** (AM.I.I.K1)
   The holder of a pilot certificate that flies that particular aircraft.

9. **Who is authorized to rebuild an aircraft engine and issue a zero-time record?** (AM.I.I.K1)
   The manufacturer of the engine or a facility approved by the manufacturer.

10. **Who is authorized to approve an aircraft for return to service after a minor alteration has been performed on the airframe?** (AM.I.I.K1)
    A certificated Aviation Mechanic holding an Airframe rating.

11. **What is the minimum age for an Aviation Mechanic certificate?** (AM.I.I.K1)
    18 years.

12. **What are the two ratings that can be issued to an Aviation Mechanic certificate?** (AM.I.I.K1)

    Airframe and Powerplant.

13. **How many months of practical experience is needed to qualify for the Aviation Mechanic certificate with both Airframe and Powerplant ratings?** (AM.I.I.K1)

    30 months.

14. **What can be used in place of the 30 months of experience to qualify to take the mechanic tests?** (AM.I.I.K1)

    A certificate of completion from a certificated aviation maintenance technician school.

15. **What type of experience is required to take the tests for Aviation Mechanic certification?** (AM.I.I.K1)

    Practical experience with the procedures, practices, materials, tools, machine tools, and equipment generally used in constructing, maintaining, or altering airframes, or powerplants appropriate to the rating sought.

16. **What tests are used to demonstrate that a mechanic applicant has the proper knowledge?** (AM.I.I.K1)

    The written knowledge tests.

17. **What tests are used to demonstrate that a mechanic applicant meets the minimum skill requirements?** (AM.I.I.K1)

    The oral and practical tests.

18. **What certificate and ratings are required for a mechanic to conduct a 100-hour inspection and approve the aircraft for return to service?** (AM.I.I.K1)

    An Aviation Mechanic certificate with both Airframe and Powerplant ratings.

19. **What certificate and ratings are required for a mechanic to conduct an annual inspection and approve the aircraft for return to service?** (AM.I.I.K1)

    An Aviation Mechanic certificate with an Inspection Authorization.

20. **When, after making a permanent change of address, is the holder of an Aviation Mechanic certificate required to notify the FAA?** (AM.I.I.K1)

    Within 30 days.

21. **Is a certificated Aviation Mechanic with Airframe and Powerplant ratings authorized to perform 100-hour and annual inspections?** (AM.I.I.K1)

    No. Only 100-hour inspections. Inspection Authorization is required to perform an annual inspection.

22. **A certificated Aviation Mechanic may not supervise the maintenance, preventive maintenance, or alteration of, or approve and return to service, any aircraft or appliance, or part thereof, for which the mechanic is rated unless he or she has met what stipulation?** (AM.I.I.K1)

    The mechanic must have satisfactorily performed the work concerned at an earlier date.

23. Is it permissible for a mechanic with an Airframe and Powerplant rating to replace the tachometer cable for a magnetic drag tachometer? (AM.I.I.K1)

Yes, this does not constitute a repair to an instrument.

24. Is a certificated Aviation Mechanic with an Airframe and Powerplant rating allowed to zero the pointers on a fuel pressure gauge? (AM.I.I.K1)

No, a mechanic is not allowed to make any repairs or alterations to an aircraft instrument.

25. How many months of experience within a 24-month period must a mechanic have to exercise the privileges of his or her certificate? (AM.I.I.K2)

At least 6 months.

26. How can a mechanic re-establish maintenance currency once it is lost? (AM.I.I.K2)

By working under supervision for 6 months, or if the Administrator has found that they are able to do the work.

27. What record must be made of the compliance of an Airworthiness Directive? (AM.I.I.K3)

An entry must be made in the aircraft maintenance records stating that the AD has been complied with. This entry must include the AD number and revision date, the date of compliance, the aircraft total time in service, the method of compliance, and whether or not this is a recurring AD. If it is a recurring AD, the time of next compliance must be noted.

28. What minimum information must be contained in a maintenance record? (AM.I.I.K3)

All maintenance entries at a minimum must include a description of the work performed, references to data acceptable to the Administrator, the date the work was completed, the name and certificate number of the person performing the work, and, if different, the signature and certificate number of the person approving the aircraft for return to service.

29. What records must be made of a 100-hour inspection before the aircraft is approved for return to service? (AM.I.I.K4)

An entry must be made in the aircraft maintenance records that describes the type, the extent, and the date of the inspection; the aircraft total time in service; and the signature, certificate type, and number of the person approving or disapproving the aircraft for return to service.

30. Where can you find an example of the correct type of write-up to use for recording a 100-hour inspection in the aircraft maintenance records? (AM.I.I.K4)

In 14 CFR §43.11.

31. What action must a mechanic take if the aircraft he is inspecting on a 100-hour inspection fails because of an unairworthy component? (AM.I.I.K4)

The aircraft maintenance records must indicate that the aircraft has been inspected and found to be in an unairworthy condition because of certain discrepancies. A signed and dated list of these discrepancies must be given to the owner or lessee of the aircraft.

32. For how long must the record of a 100-hour inspection be kept? (AM.I.I.K4)

For one year or until the next 100-hour inspection is completed.

33. What record must be made of a major repair to an aircraft structure? (AM.I.I.K5)

An FAA Form 337 must be completed for the repair, and a record must be made in the aircraft maintenance records referencing the Form 337 by its date.

34. How many copies must be made of a Form 337 after a major airframe repair? What is the disposition of each copy? (AM.I.I.K5)

At least two copies must be made. The original signed form goes to the aircraft owner, and a copy goes to the FAA district office.

35. How long must major repair or alteration records be kept? (AM.I.I.K5)

A copy of all Form 337s must be kept for the life of the aircraft.

36. What is a malfunction and defect report? (AM.I.I.K5)

Malfunction and Defect Reports (FAA Form 8010-4) are a voluntary reporting mechanism to report unusual problems or weakness not associated with expected wear.

37. Where can you find a list of the basic items that must be inspected on a 100-hour inspection? (AM.I.I.K6)

In 14 CFR Part 43, Appendix D.

38. What is meant by a progressive inspection? (AM.I.I.K6)

An inspection that is approved by the FAA FSDO in which an aircraft is inspected according to an approved schedule. This schedule allows the complete inspection to be conducted over a period of time without having to keep the aircraft out of service as long as would be necessary to perform the entire inspection at one time.

39. What is meant by permanent maintenance records? (AM.I.I.K6)

Permanent maintenance records are kept for the life of the aircraft and are transferred when a change in ownership occurs. Permanent records include: total time in service of the airframe, engines, propellers, and rotors; the status of life-limited parts; time since overhaul on items with specified time limits; current inspection status; current AD status including AD numbers, revision dates, recurrence dates, and methods of compliance; and copies of any form 337.

40. What is the purpose of 14 CFR Part 43? (AM.I.I.K6)

It describes maintenance, preventive maintenance, rebuilding, and alteration of certificated aircraft.

41. What is preventive maintenance, who may perform preventive maintenance, and what maintenance record entries are required? (AM.I.I.K6)

Preventive maintenance is limited to the list of items given in 14 CFR Part 43, Appendix A, and may not include any complex assemblies. A person holding at least a Private Pilot Certificate may approve the aircraft for return to service and must make an entry in the aircraft maintenance records.

42. What determines if an alteration or repair is considered major? (AM.I.I.K7)

The lists of the types of repairs and alterations that are considered major are listed in 14 CFR Part 43, Appendix A.

43. **What determines if an alteration or repair is considered minor?** (AM.I.I.K7)
    It is considered minor if it is not listed in 14 CFR Part 43, Appendix A, as a major repair or alteration.

44. **What is done with the aircraft maintenance records that include the current status of the applicable Airworthiness Directives when the aircraft is sold?** (AM.I.I.K9)
    These maintenance records must be transferred with the aircraft when it is sold.

45. **What is a Type Certificate Data Sheet?** (AM.I.I.K9)
    The TCDS is a formal description of the aircraft, engine, or propeller. It lists limitations and information required for type certification including airspeed limits, weight limits, thrust limitations, and so forth.

46. **What are Aircraft Specifications?** (AM.I.I.K9)
    Documents that include basically the same information as the TCDS but are issued for aircraft, engines, and propellers certificated under the Air Commerce Regulations.

47. **What document specifies the type of fuel that should be used in an airplane?** (AM.I.I.K9)
    The TCDS for that airplane.

48. **What document would you use to find the control surface movement for a specified airplane?** (AM.I.I.K9)
    The TCDS for that airplane.

49. **Why is it necessary to refer to the TCDS for an airplane when conducting a 100-hour inspection?** (AM.I.I.K9)
    The TCDS includes the specifications required for the aircraft to maintain its airworthy status.

50. **What FAA publication describes methods of nondestructive testing?** (AM.I.I.K9)
    AC 43-3, *Nondestructive Testing in Aircraft*.

51. **What is the purpose of an AD?** (AM.I.I.K9)
    It provides guidance and information to owners and operators of aircraft informing them of the discovery of a condition that prevents the aircraft from continuing to meet its conditions for airworthiness.

52. **What is the first step in the issuance of an AD?** (AM.I.I.K9)
    A notice of proposed rulemaking (NPRM) is published in the *Federal Register*.

53. **How is information on an AD disseminated?** (AM.I.I.K9)
    It is printed and distributed by first class mail to the registered owners and certain known operators of the product(s) affected.

54. **What type of AD may be adopted without an NPRM?** (AM.I.I.K9)
    ADs of an urgent nature are issued as immediately adopted rules without prior notice.

55. **How is information on an emergency AD sent to the owner or operator of an affected aircraft or other product?** (AM.I.I.K9)
By first-class mail, telegram, or other electronic method.

56. **Where can you find a list of all of the ADs that apply to aircraft, engines, propellers, or appliances?** (AM.I.I.K9)
On the FAA's website under the Dynamic Regulatory System (DRS).

57. **How can you get information on subscribing to the Airworthiness Directives?** (AM.I.I.K9)
Contact FAA, Manufacturing Standards Section (AFS-613), PO Box 26460, Oklahoma City, OK 73125-0460.

58. **What is the significance of the identification number 91-08-07 R1?** (AM.I.I.K9)
91—This AD was issued in 1991.
08—This AD was issued in the eighth biweekly period (15th or 16th week) of 1991.
07—This is the seventh AD issued during this period.
R1—This is the first revision of this AD.

59. **Is it mandatory that the information in an AC be complied with?** (AM.I.I.K9)
No, this information is advisory in nature.

60. **With which part of 14 CFR is AC 43.13-1B associated?** (AM.I.I.K9)
14 CFR Part 43.

61. **May all of the information in AC 43.13-1B be used as approved data?** (AM.I.I.K9)
No, it is acceptable, but not necessarily approved data.

62. **What is the purpose of General Aviation Airworthiness Alerts?** (AM.I.I.K9)
They contain information gleaned from Malfunction and Defect Reports to warn maintenance personnel of problems that have been reported.

63. **If an individual develops a different process for complying with an AD, are they allowed to use that process?** (AM.I.I.K10)
They are not allowed to deviate from the AD unless they submit the process to the Administrator as an Alternative Method of Compliance (AMOC), and the AMOC is approved.

64. **Who issues a service bulletin and what is its purpose?** (AM.I.I.K11)
It is issued by the manufacturer and used to outline procedures to make the product operate more efficiently and/or safer.

65. **May the data in an aircraft maintenance manual be used as approved data for an aircraft repair?** (AM.I.I.K11)
No. An aircraft maintenance manual contains repair data acceptable to the FAA for minor repairs only. If approved data is needed for a major repair, consult the aircraft manufacturer's FAA-approved Structural Repair Manual, an FAA DER (Designated Engineering Representative), or in some cases, AC43.13-1.

66. **Where would you find the dimensional tolerances for the wrist pin fit in an aircraft engine?** (AM.I.I.K11)
In the overhaul manual for that engine.

67. **What document would you use to find an approved repair for a damaged wing spar?** (AM.I.I.K11)
The structural repair manual for the aircraft.

68. **What document would you use to find the part number for a landing light bulb for an airplane?** (AM.I.I.K11)
The illustrated parts catalog.

69. **What document would you use to troubleshoot a malfunctioning electrical flap system?** (AM.I.I.K11)
The aircraft wiring diagram manual.

70. **Where would you find the empty weight of an airplane?** (AM.I.I.K11)
In the aircraft weight and balance records.

71. **What is the purpose of a minimum equipment list (MEL)?** (AM.I.I.K11)
An MEL permits operations with certain inoperative items of equipment for the minimum period of time necessary until repair or replacement can be accomplished.

72. **What is a master minimum equipment list (MMEL)?** (AM.I.I.K11)
A document approved by the FAA that lists the minimum operative instruments and equipment required for safe flight in that aircraft type in each authorized operating environment.

73. **Where are the FAA databases that include advisory circulars, ADs, TCDSs, and supplemental type certificates (STCs)?** (AM.I.I.K12)
On the FAA's Dynamic Regulatory System (DRS) website.

74. **Where does an aircraft manufacturer specify the compliance requirements for methods, techniques, and practices?** (AM.I.I.K13)
In the Maintenance Manual and Instructions for Continued Airworthiness.

75. **Where does an aircraft manufacturer specify maintenance and inspection intervals?** (AM.I.I.K14)
In the Maintenance Manual and Instructions for Continued Airworthiness.

76. **Where does an aircraft mechanic find FAA-approved maintenance data, including maintenance manuals and other methods, techniques, and practices acceptable to the Administrator?** (AM.I.I.K15)
In the aircraft manufacturer's maintenance manual and Instructions for Continued Airworthiness, and in advisory circulars (including AC 43.13-1 and AC 43.13-2).

77. **What are the differences between approved data and acceptable data?** (AM.I.I.K16)
FAA "approved data" is required for major repairs and alterations, and "acceptable data" is required for minor repairs and alterations.

78. **What are "Instructions for Continued Airworthiness"?** (AM.I.I.K17)
FAA Order 8110.54A, Appendix I, defines Instructions for Continued Airworthiness (ICA) as documentation that gives instructions and requirements for the maintenance that is essential to the continued airworthiness of an aircraft, engine, or propeller, including appliances, components, or modifications.

79. **What is an airplane flight manual (AFM) or pilot's operating handbook (POH)?** (AM.I.I.K17)
Required documents that must be carried in an aircraft that detail the operating limitations specific to that aircraft and its engines.

80. **Where are the definitions found for *Warnings*, *Cautions*, and *Notes* that are used in maintenance and operating manuals?** (AM.I.I.K18)
The definitions for these words are found at the beginning of the manuals that they are used in, usually in Chapter 1.

81. **What FAA regulation provides guidance for the use of placards when operating with inoperative equipment?** (AM.I.I.K20)
14 CFR §91.213.

82. **How are parts identified in a manufacturer's parts manual to specify the specific serial number of aircraft that the part is valid for?** (AM.I.I.K21)
Through the use of "Useable On" or effectivity codes.

83. **Where are the part numbers found for a specific aircraft model and serial number?** (AM.I.I.K22)
In the manufacturer's parts manual or Illustrated Parts Catalog (IPC).

84. **How soon after a change in permanent mailing address should a mechanic notify the FAA?** (AM.I.I.K23)
Within 30 days.

85. **What is the method of notifying the FAA of a change of address?** (AM.I.I.K23)
Either through the mail with the FAA's office in Oklahoma City, or online through the FAA's Airmen Services.

Risk Management

1. **Why is it important for maintenance documentation to be accurate or complete?** (AM.I.I.R1)
Incomplete or inaccurate aircraft records lead to incorrect maintenance history and understanding of the status of the aircraft, which can cause unairworthy and possibly unsafe conditions.

2. **Why is it important to use SDSs (Safety Data Sheets)?** (AM.I.I.R2)
To promote safe working conditions for personnel and prevent damage to aircraft components caused by the improper use of materials and chemicals.

3. **Explain the dangers of complacency during the documentation phase of maintenance procedures.** (AM.I.I.R3)

   Complacency during the documentation phase can cause incomplete records and uncertainty of aircraft and component status.

4. **Why is it important to follow warnings, cautions, and notes in maintenance and operational manuals?** (AM.I.I.R4)

   Failure to follow warnings, cautions, and notes can lead to unsafe conditions and damage to the aircraft, aircraft components, and other aircraft and injury to personnel.

5. **Describe how to determine if a part applies to a given aircraft.** (AM.I.I.R5)

   When looking up a part in the aircraft's Illustrated Parts Catalog (IPC), it is important to understand aircraft configuration as listed in the aircraft Type Certificate Data Sheet (TCDS) and to know the specific serial number of the aircraft. In addition, the aircraft records should be checked for approved modifications or repairs as shown in AD records and in Supplemental Type Certificates (STC) in the aircraft maintenance records.

Skills

1. **Prepare a Form 337 describing a major repair or major alteration that is specified by the examiner.** (AM.I.I.S1)

2. **Given a list of repairs and alterations to an aircraft and engine, identify the operations that require a Form 337 to be filled out.** (AM.I.I.S1)

3. **Complete a Form 337 using information provided.** (AM.I.I.S1)

4. **Examine an FAA Form 337 provided by the examiner and determine if the form was filled out accurately.** (AM.I.I.S2)

5. **Explain to the examiner who is responsible for assuring that all the required maintenance is done on an aircraft.** (AM.I.I.S3)

6. **Demonstrate to the examiner the part of the Federal Regulations that describes the maintenance required for certificated aircraft.** (AM.I.I.S3)

7. **Explain to the examiner when a flight test is required after maintenance.** (AM.I.I.S3)

8. **Explain to the examiner which aircraft require a 100-hour inspection and which ones require an annual inspection.** (AM.I.I.S3)

9. **Prepare a maintenance record entry that records the proper compliance with an Airworthiness Directive specified by the examiner.** (AM.I.I.S4)

10. **Prepare a maintenance record entry that records preventive maintenance done on an aircraft specified by the examiner.** (AM.I.I.S4)

11. **Create a current equipment list for an aircraft, listing all equipment installed.** (AM.I.I.S5)

12. **Using a copy of an MEL furnished by the examiner, show what instruments are allowed to be inoperative.** (AM.I.I.S5)

13. **Demonstrate your ability to locate the following data from a Type Certificate Data Sheet for an aircraft specified by the examiner:**
    a. **Maximum allowable gross weight**
    b. **The never-exceed airspeed**
    c. **The minimum grade of fuel allowed**
    d. **The center of gravity limits**
    e. **Flight control travel limits**
    f. **The conformity of an aircraft instrument range markings and/or placards**
    g. **Approved tires for installation**

    (AM.I.I.S6)

14. **Demonstrate to the examiner the way to locate the required service bulletins that apply to a specific aircraft or engine.** (AM.I.I.S6)

15. **Write up a maintenance record of a 100-hour inspection on an aircraft specified by the examiner.** (AM.I.I.S7)

16. **Prepare a maintenance record entry that records the required tests and inspection for an altimeter system installed on an aircraft flown under instrument flight rules.** (AM.I.I.S7)

17. **Make a proper maintenance record entry for the adjustment of the wing flap indicating system.** (AM.I.I.S7)

18. **Locate an AD specified by the examiner for a particular aircraft and answer these questions:**
    a. **What is the effective date of this AD?**
    b. **What models of aircraft are affected by this AD?**
    c. **What serial numbers are affected by this AD?**
    d. **How soon must this AD be complied with?**

    (AM.I.I.S8)

19. **Prepare a master AD list for a specific airframe, engine, and/or propeller and determine applicability by make, model, and serial number.** (AM.I.I.S8)

20. **Check a technical standard order (TSO) part for proper markings.** (AM.I.I.S9)

21. **Use an illustrated parts catalog to find the part number of an O-ring specified by the examiner.** (AM.I.I.S10)

22. **Locate the applicable supplemental type certificates for a given aircraft.** (AM.I.I.S11)

23. **Determine the conformity of aircraft instrument range markings and placards in an aircraft flight deck.** (AM.I.I.S12)

24. **Demonstrate your ability to use a structural repair manual to design a repair to a damaged wing spar.** (AM.I.I.S13)

25. **Determine the maximum allowable weight of a given aircraft.** (AM.I.I.S14)

26. **Determine if a given repair or alteration is minor or major.** (AM.I.I.S15)

27. **Using AC 43.13-1B, design a repair for a damaged bulb angle stringer in the upper surface of a wing.** (AM.I.I.S16)

28. **Explain to the examiner the difference between "approved data" used for a major repair or alteration, and "acceptable data" used for a minor repair or alteration.** (AM.I.I.S17)

29. **Prepare a maintenance record entry that approves an aircraft for return to service after a 100-hour inspection. Include a list of some allowable inoperative instruments or equipment.** (AM.I.I.S18)

30. **Prepare a maintenance record entry that disapproves an aircraft for return to service after a 100-hour inspection.** (AM.I.I.S18)

J. Physics for Aviation

References: AC 43.13-1; FAA-H-8083-30

Knowledge

1. **What is meant by matter?** (AM.I.J.K1)
   Anything that occupies space and has weight.

2. **What are the three basic physical states in which matter can exist?** (AM.I.J.K1)
   Solid, liquid, and gas.

3. **What is meant by energy?** (AM.I.J.K1)
   The capacity for doing work.

4. **What formula is used to find the amount of work done when an object is moved across a floor?** (AM.I.J.K2)
   Work = Force × Distance

5. **What determines the mechanical advantage of an arrangement of ropes and pulleys?** (AM.I.J.K2)
   The number of ropes that support the weight.

6. **What determines the mechanical advantage of a gear train?** (AM.I.J.K2)
The ratio between the number of teeth on the drive gear and the number of teeth on the driven gear.

7. **What is meant by the fulcrum of a lever?** (AM.I.J.K3)
The point about which the lever rotates.

8. **What are examples of a first-class lever, a second-class lever, and a third-class lever?** (AM.I.J.K3)
First-class: A screwdriver being used to pry the lid from a can of paint.
Second-class: A wheelbarrow.
Third-class: A hydraulically retracted landing gear.

9. **What is meant by pressure?** (AM.I.J.K4)
Force that acts on a unit of area.

10. **What is the standard sea level atmospheric pressure expressed in inches of mercury and in pounds per square inch?** (AM.I.J.K4)
29.92 inches of mercury and 14.69 pounds per square inch.

11. **What characteristic of the atmosphere determines the speed of sound?** (AM.I.J.K4)
Its temperature.

12. **What is meant by a temperature of absolute zero?** (AM.I.J.K4)
The temperature at which all molecular movement stops.

13. **What is the Celsius equivalent of a temperature of 50°F?** (AM.I.J.K4)
10°C

14. **What are three methods of heat transfer?** (AM.I.J.K4)
Conduction, convection, and radiation.

15. **What is meant by the absolute humidity of the atmosphere?** (AM.I.J.K4)
The actual amount of water that is in a given volume of air.

16. **What causes ice to change into liquid water?** (AM.I.J.K4)
The absorption of heat energy.

17. **What happens inside a solid material when it absorbs heat energy?** (AM.I.J.K4)
The molecules within the material move faster.

18. **What is the basic unit of heat in the English system?** (AM.I.J.K4)
British Thermal Unit.

19. **How much work will one Btu of heat energy perform?** (AM.I.J.K4)
778 foot-pounds of work.

20. **What is the basic unit of heat in the Metric system?** (AM.I.J.K4)
Calorie.

21. **How much heat energy is in a small calorie?** (AM.I.J.K4)
The amount of heat energy that will raise the temperature of 1 gram of water 1°C.

22. **How much heat energy is in a large calorie (Calorie)?** (AM.I.J.K4)
The amount of heat energy that will raise the temperature of 1 kilogram of water 1°C.

23. **What is the first law of thermodynamics?** (AM.I.J.K4)
Heat energy can neither be created nor destroyed, it can only be changed in its form.

24. **What is the second law of thermodynamics?** (AM.I.J.K4)
Heat energy can only flow from a body having a high temperature to a body having a lower temperature.

25. **What is an example of heat transfer by conduction?** (AM.I.J.K4)
Removal of heat from an engine cylinder by air flowing over its surface.

26. **What is an example of heat transfer by convection?** (AM.I.J.K4)
The uniform heating of the air in a room by a floor heater. The heated air rises and forces the cooler air down so it can be heated by conduction.

27. **What is an example of heat transfer by radiation?** (AM.I.J.K4)
The heating of the Earth's surface by heat transmitted through space from the sun.

28. **Why do most metals expand when they are heated?** (AM.I.J.K4)
As heat is absorbed, the electrons move faster and expand their orbits in the molecules of the metal.

29. **What is Bernoulli's principle?** (AM.I.J.K5)
Bernoulli's principle states that an increase in speed of a fluid results in a decrease in pressure of the same fluid.

30. **What is Newton's first law of motion?** (AM.I.J.K6)
Newton's first law of motion states that a body at rest remains at rest, and a body in motion will continue in a straight line unless acted on by a force.

31. **What are two types of fluids?** (AM.I.J.K7)
Liquid and gaseous.

32. **What is meant by the density of a fluid?** (AM.I.J.K7)
The mass-per-unit volume of the fluid.

33. **What is meant by the specific gravity of a fluid?** (AM.I.J.K7)
The ratio of the density of the fluid to the density of pure water.

34. **What effect does the increase in temperature of a confined gas have on its pressure?** (AM.I.J.K7)
When the volume of a gas remains constant, an increase in its temperature increases its pressure.

35. What effect does high humidity have on piston engine performance? (AM.I.J.K7)

Water vapor is less dense than dry air and thus high humidity decreases the density of the air. The less dense air decreases engine performance.

36. How many cubic inches of fluid is forced out of a cylinder by a piston with an area of 20 square inches, when the piston moves five inches? (AM.I.J.K7)

100 cubic inches.

37. What will happen to the pressure of a confined gas if the temperature of the gas is increased? (AM.I.J.K7)

The pressure will increase.

38. What is meant by the density of air? (AM.I.J.K8)

The weight of a given volume of air.

39. What is meant by relative wind with regard to an airfoil? (AM.I.J.K8)

The direction the wind strikes an airfoil.

40. What is meant by the mean aerodynamic chord (MAC)? (AM.I.J.K8)

MAC is the average distance from the leading edge to the trailing edge of the wing.

41. What is meant by the angle of attack? (AM.I.J.K8)

The acute angle formed between the chord line of an airfoil and the direction the air strikes the airfoil.

42. What is meant by the critical angle of attack? (AM.I.J.K8)

The highest angle of attack at which the air passes over the airfoil in a smooth flow. Above the critical angle of attack the airflow breaks away and becomes turbulent.

43. What is meant by the stagnation point of an airfoil? (AM.I.J.K8)

The point on the leading edge of an airfoil at which the airflow separates, some flowing over the top and some over the bottom.

44. What is the difference between speed and velocity? (AM.I.J.K8)

Speed is the rate at which an object is moving. Velocity is the vector quantity that expresses both the rate and direction an object is moving.

45. What is meant by air density? (AM.I.J.K8)

The mass of air in a given volume.

46. What is meant by weight? (AM.I.J.K8)

The measure of the force of gravity acting on a body.

47. What is meant by thrust? (AM.I.J.K8)

The forward aerodynamic force produced by a propeller, fan, or turbojet engine as it forces a mass of air to the rear, behind the airplane.

48. What is meant by drag? (AM.I.J.K8)

The aerodynamic force acting in the same plane as the relative wind striking an airfoil. Drag acts in the direction opposite to that of thrust.

49. What is meant by autorotation in a helicopter? (AM.I.J.K8)

The aerodynamic force that causes a helicopter rotor to spin with no engine power applied to the rotor system.

50. What is meant by dissymmetry of lift produced by a helicopter rotor? (AM.I.J.K8)

The difference in lift between the two sides of the rotor disc when the helicopter is in forward flight. The side with the advancing blade produces the greater lift because the forward speed adds to the rotor speed. The side with the retreating blade produces less lift because the forward speed subtracts from the rotor speed.

51. What is meant by a blade stall of a helicopter rotor? (AM.I.J.K8)

A condition of flight in which the retreating blade is operating at an angle of attack higher than will allow for the air to flow over its upper surface without turbulence.

52. What is meant by translational lift in a helicopter? (AM.I.J.K8)

The additional lift produced by a helicopter rotor as the helicopter changes from hovering to forward flight.

53. What is meant by ground effect in helicopter flight? (AM.I.J.K8)

An increase in lift when a helicopter is flying at an altitude of less than half the rotor span. This increase is produced by the effective increase in the angle of attack caused by the deflection of the downwashed air.

54. What is meant by ground resonance in a helicopter? (AM.I.J.K8)

The destructive vibration that occurs when the helicopter touches down roughly and unevenly. The shock throws a load into the lead-lag hinges of the rotor blades and causes them to oscillate about this hinge. If the frequency of this oscillation is the same as the resonant frequency of the fuselage, the energy will cause the helicopter to strike the ground hard with the opposite skid or wheel. If corrective action is not taken immediately, ground resonance can destroy the helicopter.

55. What effect does an increase in density altitude have on engine performance? (AM.I.J.K9)

As density altitude increases, air density decreases and engine performance decreases.

56. How do flaps allow an airplane to take off and land at a slow airspeed? (AM.I.J.K10)

The extension of flaps increases the camber of the airfoil, causing greater lift and drag.

57. Why do slots and slats delay the onset of a stall to a higher angle of attack? (AM.I.J.K10)

They allow the airflow to remain attached to the upper surface of the wing at higher angles of attack, reducing stall speed.

58. What is the purpose of stall strips on the leading edge of an airplane wing? (AM.I.J.K11)

Stall strips cause a disruption to the airflow as angle of attack increases, causing the inboard section of the wing to stall before the outboard wing.

59. **What is the function of wing fences on aircraft?** (AM.I.J.K11)
Wing fences, or stall fences, delay the stall from reaching the ailerons, allowing improved aileron control at slower airspeeds.

60. **How do vortex generators prevent the onset of airflow separation at high angles of attack?** (AM.I.J.K11)
Vortex generators create a small vortex, or disturbance that causes the air to remain attached to the surface at higher angles of attack.

61. **What is the relationship between temperature, density, weight, and volume?** (AM.I.J.K12)
As temperature increases, density decreases and volume increases. As density decreases, the weight of a set volume also decreases.

62. **What effect does an increase in the volume of a gas have on its temperature if its pressure remains constant?** (AM.I.J.K13)
As the volume of a gas increases with a constant pressure, the temperature decreases.

63. **What effect does an increase in the pressure of a confined gas have on its temperature?** (AM.I.J.K13)
Increasing the pressure of a confined gas increases its temperature.

64. **What is the difference in the fluids used in a hydraulic system and those used in a pneumatic system?** (AM.I.J.K13)
Fluid used in a hydraulic system is incompressible. Fluid used in a pneumatic system is compressible.

65. **What effect on density altitude is caused by an increase in air temperature?** (AM.I.J.K13)
As the temperature increases, the air density decreases and the density altitude increases.

66. **What is meant by the resonant frequency of an aircraft structure?** (AM.I.J.K13)
The frequency that produces the greatest amplitude of vibration in the structure.

67. **How much force is produced by 1,000 psi of hydraulic pressure acting on a piston with an area of 20 square inches?** (AM.I.J.K13)
20,000 pounds.

Risk Management

1. **What risks are associated with aircraft operations at higher density altitudes?** (AM.I.J.R1)
As density altitude increases, engine performance decreases and aircraft aerodynamic performance decreases.

2. **Explain the potential risks associated with a repair to a control surface.** (AM.I.J.R2)
An incorrectly repaired control surface can begin to flutter during flight and separate from the aircraft.

3. **What are the risks of an individual creating their own performance and testing data?** (AM.I.J.R3)

   Performance and testing data are used by aircraft manufacturers to show that their products meet the regulations that they are being certified to. The certification criteria can be complex and specific to certain aircraft and components. It is dangerous for a certified mechanic or pilot to develop their own testing of a modification without consulting the FAA Aircraft Certification Office in their district.

4. **Describe the risks associated with confusing Celsius vs. Fahrenheit, gallons vs. liters, or pounds vs. kilograms.** (AM.I.J.R4)

   Confusing English and metric units of measure can lead to misreading of performance charts, fueling an aircraft with less fuel than necessary, overloading an aircraft beyond maximum allowed weight, and many other dangers.

Skills

1. **Convert a list of Celsius temperatures into Fahrenheit temperatures.** (AM.I.J.S1)

2. **Convert a list of Fahrenheit temperatures into Celsius temperatures.** (AM.I.J.S1)

3. **Determine the atmospheric density altitude.** (AM.I.J.S2)

4. **Determine the atmospheric pressure altitude.** (AM.I.J.S3)

5. **Using a dimensioned diagram of a hydraulic cylinder furnished by the examiner, find the amount of force exerted by a specified amount of hydraulic pressure.** (AM.I.J.S4)

6. **Demonstrate how different types of levers create a mechanical advantage.** (AM.I.J.S5)

7. **From a drawing of a lever furnished by the examiner, identify the class of lever and its mechanical advantage.** (AM.I.J.S5)

8. **Find the amount of force needed to roll a barrel of oil up an inclined plane when the length of the plane, the weight of the barrel, and the height the barrel is raised are all known.** (AM.I.J.S5)

9. **Using a diagram of a gear train furnished by the examiner, find the speed and direction of rotation of the output shaft when the speed and direction of the input shaft are known. Compute the mechanical advantage of the gear train.** (AM.I.J.S5)

10. **Design a mechanical pulley system.** (AM.I.J.S5)

11. **Draw an inclined plane on paper, indicating the mechanical advantage.** (AM.I.J.S6)

12. **Identify the changes in pressure and velocity of air flowing through a given venturi.** (AM.I.J.S7)

13. **Explain to the examiner the way air flowing through the venturi of a float carburetor causes fuel to be drawn from the float bowl.** (AM.I.J.S7)

14. **Determine the engine horsepower for a given weight, distance, and time.** (AM.I.J.S8)

K. Inspection Concepts and Techniques

References: AC 43.13-1; FAA-H-8083-30

Knowledge

1. **What are different types of micrometer calipers?** (AM.I.K.K1)
   Outside micrometer, inside micrometer, depth micrometer, and thread micrometer.

2. **How often should torque wrenches be calibrated?** (AM.I.K.K2)
   Depending on the tool, some are calibrated every 6 months, while others are calibrated every 12 months.

3. **What is the proper type of nondestructive inspection to use for locating surface cracks in an aluminum alloy casting or forging?** (AM.I.K.K3)
   Fluorescent or visual dye penetrant.

4. **What is the principle upon which ultrasonic inspection is based?** (AM.I.K.K3)
   Any fault within a material will change the material's resonant frequency. Comparing the resonant frequency of a known sound material with the material under test will indicate the presence of a fault.

5. **What is the procedure when making a dye penetrant inspection of a part?** (AM.I.K.K3)
   Clean the part thoroughly, apply the penetrant, and allow it to soak for the recommended dwell time. Remove all of the penetrant from the surface and apply the developer.

6. **In addition to 100-hour and annual inspections, where would an aircraft operator find the approved inspection programs for an aircraft?** (AM.I.K.K4)
   In the aircraft manufacturer's maintenance manual and Instructions for Continued Airworthiness (ICA).

7. **What inspection method would be most appropriate for checking a nonferrous metal part for intergranular corrosion?** (AM.I.K.K5)
   Eddy current inspection.

8. **What inspection method would be most appropriate for checking the internal structure of an airplane wing for corrosion?** (AM.I.K.K5)
   X-ray inspection.

9. **What inspection method would be most appropriate for checking a steel landing gear component for stress cracks?** (AM.I.K.K5)
   Magnetic particle inspection.

10. **What type of inspection is best suited for detecting a fault within a piece of nonferrous metal?** (AM.I.K.K5)
    Ultrasonic inspection.

11. **Why is it important that all parts be thoroughly cleaned before they are inspected by the dye penetrant method?** (AM.I.K.K5)
    Any grease or dirt in a fault will keep the penetrant from seeping into the fault.

12. **What is the procedure to use when performing a magnetic particle inspection of a part?** (AM.I.K.K5)
    Thoroughly clean the part, magnetize it as directed by the appropriate service manual, flow the indicating medium over the surface, and inspect it under a "black" light. When the inspection is complete, thoroughly demagnetize the part.

13. **Why is it necessary to magnetize a part both circularly and longitudinally when inspecting a steel part by the magnetic particle method?** (AM.I.K.K5)
    To detect faults that extend across as well as lengthwise within the part.

14. **Does circular magnetization detect faults that are across or lengthwise to the part?** (AM.I.K.K5)
    Lengthwise.

15. **Why is it important that all engine parts that have been inspected by the magnetic particle method be completely demagnetized?** (AM.I.K.K5)
    If the parts are not completely demagnetized, they will attract steel particles that are produced by engine wear and will cause damage to bearing surfaces.

16. **What is the principle of eddy current inspection?** (AM.I.K.K5)
    A current is induced into the metal being tested by a test probe. The amount of induced current is determined by the conductivity, mass, and permeability of the material, and the presence of any faults or voids.

Risk Management

1. **Why is it important to demagnetize a component following a magnetic particle inspection?** (AM.I.K.R1)
   To prevent magnetized parts from attracting filings, grindings, chips, or other steel particles, which can cause scoring, damage, and premature wear.

2. **What are the dangers of the incorrect use of precision micrometers and calipers?** (AM.I.K.R2)
   Incorrect measurement of close tolerance components can lead to failure of critical parts and assemblies.

3. **Why is calibration of precision measuring equipment important?** (AM.I.K.R3)
   Measuring equipment that is out of calibration can result in incorrect measurements, which can lead to component and system failure.

4. **What are the risks of not using the correct inspection techniques?** (AM.I.K.R4)
   Incorrect inspection techniques will lead to the aircraft or component not being inspected in accordance with the manufacturer's specifications, which can lead to failure of the inspected item.

5. **What precautions should be taken when using a multimeter to measure resistance, voltage, and current?** (AM.I.K.R5)
   Incorrect use of a multimeter can cause damage to the multimeter and/or result in incorrect conclusions regarding the circuit or component being tested. When used as an ohmmeter, the component being measured must be isolated (disconnected) from the circuit. If the multimeter is set for measuring current, and the leads are placed across any voltage drop, the meter can be damaged. A thorough understanding of the circuit being tested is critical for proper troubleshooting.

Skills

1. **Demonstrate how to use vernier calipers by measuring the components provided by the examiner.** (AM.I.K.S1)

2. **Demonstrate how to use a micrometer by measuring the components provided by the examiner.** (AM.I.K.S2)

3. **Demonstrate how to measure the bore of a cylinder using gauges and micrometers.** (AM.I.K.S3)

4. **Perform a visual inspection of the component provided by the examiner and explain the proper procedures and tools necessary to complete the inspection.** (AM.I.K.S4)

5. **Using dye penetrant, inspect a part furnished by the examiner.** (AM.I.K.S5)

6. **Perform the inspection required by an AD for the aircraft or component provided by the examiner.** (AM.I.K.S6)

7. **Inspect a part by the magnetic particle inspection method. Correctly magnetize the part, inspect it, and properly demagnetize it.** (AM.I.K.S7)

8. **Using eddy current inspection equipment, inspect a part furnished by the examiner.** (AM.I.K.S7)

9. **Using the information furnished by the examiner, find the magnetizing current recommended for magnetically inspecting a part specified by the examiner.** (AM.I.K.S7)

10. **Demonstrate how to perform a tap test on the composite component provided by the examiner.** (AM.I.K.S8)

L. Human Factors

References: AC 43.13-1; FAA-H-8083-30

Knowledge

1. **How would you describe an organization that has a positive safety culture?** (AM.I.L.K1)

   An organization that has a set of shared beliefs and attitudes that promote safety through communication, training, and other actions within all levels of the workforce.

2. **What are four areas that are assessed in an organization when evaluating human factors relating to safety?** (AM.I.L.K1)

   The PEAR model describes these areas as people (the workforce), the environment where people work, the actions people take and the resources that are available.

3. **What is a safety management system (SMS) in an organization?** (AM.I.L.K1)

   An SMS is a systematic approach to providing acceptable levels of safety risk within the organization. This approach ties together safety policies, risk management, safety assurance, and safety promotion.

4. **What are the different types of human error?** (AM.I.L.K2)

   Human errors can be grouped omission, commission, and extraneous errors. Omission errors are tasks which should have been performed but were not. Commission errors are tasks which were done incorrectly. Extraneous errors occur when an individual performs a task that was not authorized.

5. **What is the difference between an active error and a latent error?** (AM.I.L.K2)

   An active error is an error that is committed by the person doing the work. A latent error is one that comes from an administrative decision, a flaw in the system, or a culture that exists in a workplace. Latent errors may take a long time to be seen and usually show up as an active error that has its roots in the latent error.

6. **What are the different levels of consequences of human error?** (AM.I.L.K2)

   - Little or no effect.
   - Damage to equipment or hardware. The equipment or hardware can be replaced, but at a loss of time and financial setback.
   - Personal injury. Injury to an aviation mechanic, aircraft operator, or passenger.
   - Catastrophic damage (loss of life or loss of an aircraft).

7. **How can the MEDA model be used for evaluating incidents or accidents caused by an aviation mechanic or inspector?** (AM.I.L.K3)

   The MEDA model is used to identify contributing factors behind the error. MEDA stands for the Maintenance Error/Event Decision Aid model. The contributing factors are analyzed to determine the probability that they will cause an error, and then the error is analyzed to determine the probability that the error will cause an event.

8. **What is the SHEL model for event investigation?** (AM.I.L.K3)

   The SHEL model stands for:

   - **S**oftware/procedures—Software refers to the intangible part of the system such as rules, regulations, customs, habits, and the procedures involved.
   - **H**ardware—Hardware represents the physical aspects that the aviation mechanic works with, including the aircraft, tools, hardware, and facilities.
   - **E**nvironment—The physical environment in which the work is being done.
   - **L**iveware/personnel—Liveware is the human element in the system.

   All four aspects overlap to some extent, especially the human element. Each element is examined for interactions with the other elements when investigating the issue.

9. **How can human limitations affect an individual's performance?** (AM.I.L.K4)

   "People" is the P in the PEAR model and describes how people who do the work are the center of all work. There are many factors that can affect or limit an individual's ability to perform work. Some of these limitations can include:

   - Physical—physical size, age, strength, sensory limitations
   - Psychological—nutritional factors, health, lifestyle, fatigue, chemical dependency
   - Physiological—workload, experience, knowledge, training, attitude, mental or emotional state
   - Psychosocial—interpersonal conflicts

10. **What elements of the physical environment can affect an aviation mechanic's ability to perform work?** (AM.I.L.K5)

    An aviation mechanic's ability to perform work safely can be affected by:

    - Weather (hot, cold, rain, lightning, wind)
    - Location (protection from the elements)
    - Workspace (having enough room to adequately perform the work)
    - Shift (late nights, or being too tired, can affect an individual's capacity to think clearly and reduce manual and cognitive skills)
    - Lighting (the environment can be too dark, or too bright, to effectively see and perform work)
    - Sound level (an environment that is too loud can cause injury to hearing and disrupts the work at hand)
    - Safety (safety can be compromised by working at heights, moving equipment, hazardous materials, or any number of conditions within a work environment)

11. **How can organizational and social environments affect an aviation mechanic's ability to effectively perform work?** (AM.I.L.K5)

    The culture within which an individual works can have a large impact on their ability to work safely and effectively. Areas that impact an organization's environment are:

    - Personnel
    - Supervision
    - Labor-management relations
    - Pressures
    - Crew structures
    - Size of company
    - Profitability
    - Morale
    - Corporate culture

12. How does the reporting of hazards within an organization tie into the overall safety policy? (AM.I.L.K6)

Management of an organization defines and controls the safety policy as they are the ones who are ultimately responsible for safety within the organization. There must be clear goals that are openly communicated regarding the reporting of hazards for safety goals to be met.

13. Why is good communication important in an aviation work environment? (AM.I.L.K6)

Individuals have differing ways of expressing and receiving ideas, opinions, and instructions. If a team of aviation mechanics is going to work together on a large project, it is vital that they understand each other and can perform the work safely and efficiently. In addition to communicating clearly with each other, they must also be able to communicate clearly with supervisors, the aircraft owner, and the suppliers of aircraft parts and services.

14. What is a reactive approach to hazard identification? (AM.I.L.K6)

The reactive approach to hazard identification investigates accidents, incidents, and events. The investigation occurs after the occurrence.

15. What is a proactive approach to hazard identification? (AM.I.L.K6)

The proactive approach to hazard identification actively identifies safety hazards through analysis of the organization's activities through mandatory and voluntary reporting systems, safety audits, and safety surveys.

16. What is a predictive approach to hazard identification? (AM.I.L.K6)

The predictive approach to hazard identification captures system performance as it happens during normal operations, such as observing a mechanic's performance during a heavy check.

17. What are the elements of good teamwork? (AM.I.L.K7)

Most aviation mechanics work as part of a team to do their work. The team must function cohesively to accomplish the work at hand safely and correctly. This includes good communication, understanding the end goal, watching out for teammates, and cooperating with each other. Part of the team attitude is understanding that your actions and attitudes directly affect those around you.

18. How does leadership affect the safety culture in an organization? (AM.I.L.K7)

For a safety management system to be effective, senior management must lead by example and promote safety culture through policies, promotion, and risk management. If the leaders of a company are not completely supportive of a safety culture, it will show through, and will become less important to the rest of the company.

19. How can communication affect and facilitate teamwork? (AM.I.L.K7)

Communication is key for individuals to work together as a team. Discuss specific duties when jobs require more than one person to eliminate missing steps or tasks. When assigning work to an individual, never assume that they have all the information and resources needed for the task. Verify their understanding of critical tasks by making sure they have access to the correct manuals and understand the process to be completed.

20. What does professionalism mean from the standpoint of an aviation mechanic? (AM.I.L.K8)

Professionalism is the conduct and practice of doing one's job with skill, competence, ethics, and courtesy. A professional aviation mechanic will understand and carry out proper, recommended maintenance procedures, and will have the necessary skills to perform the task. The individual recognizes when their understanding of a task is lacking and will seek appropriate guidance. A professional will treat all other individuals with courtesy and respect, regardless of their position.

21. What does it mean for an aviation mechanic to work with integrity? (AM.I.L.K8)

To work with integrity means that the aviation mechanic will not compromise safety, airworthiness, regulations, or company policies, regardless of who sees (or does not see) what they do.

22. What does an aviation mechanic's signature on a document indicate? (AM.I.L.K8)

An aviation mechanic's signature is their word indicating that each and every task was performed as required, to the best of their ability. The signature is directly linked to the aviation mechanic's professionalism and integrity. To sign a document knowing that something has not been completed or is wrong, means that integrity has been compromised. Once compromised, it is very difficult to earn integrity back.

23. What are the procedures necessary for an effective shift change? (AM.I.L.K9)

Properly use logbooks and worksheets to communicate work accomplished and work to be done. At shift changes, maintenance personnel must discuss what work has been accomplished and what needs to be completed. Never assume work has been completed if there is no record of completion. In addition to a written record of work accomplished, there should be a verbal brief between shifts to provide further details and to allow for questions regarding the status of the work.

24. What should be included in a written transfer of a critical task? (AM.I.L.K9)

When transferring a critical task, it is important to record exactly what has been accomplished up to that point, as well as critical tasks not yet completed.

25. How does our recognition of conditions or preconditions for unsafe acts affect safety? (AM.I.L.K10)

If aviation mechanics can recognize they are in a state in which errors can occur, they can change the condition and prevent errors.

26. What is the danger of having two or more unsafe conditions present? (AM.I.L.K10)

When two or more factors are present, the error probability increases dramatically.

27. What are the "Dirty Dozen" of aviation maintenance? (AM.I.L.K10)

The "Dirty Dozen" is a list of twelve factors or conditions that could lead to an aviation mechanic making an error. The twelve factors are:

1. Lack of communication
2. Complacency
3. Lack of knowledge
4. Distractions
5. Lack of teamwork
6. Fatigue

7. Lack of resources
8. Pressure
9. Lack of assertiveness
10. Stress
11. Lack of awareness
12. Norms

28. What is an error of omission? (AM.I.L.K11)

Omission errors are tasks which should have been performed but were not. Examples are missing a step in an inspection, or not installing a washer or cotter pin.

29. What is an error of commission? (AM.I.L.K11)

Commission errors are tasks which were done incorrectly. This could be torquing a fastener to the wrong torque or timing a magneto to an engine at the wrong number of degrees.

30. What are extraneous errors? (AM.I.L.K11)

Extraneous errors occur when an individual performs a task that was not authorized.

31. What is the difference between a human error and a violation? (AM.I.L.K11)

A human error is an action performed by an individual that has unintended consequences. An action that is performed with full knowledge of what consequences are expected or possible is called a violation.

Risk Management

1. Describe the effect of reporting only select hazards. (AM.I.L.R1)

The selective reporting of hazards means that some hazards will be overlooked and not reported. This can lead to risks in critical areas being overlooked and could eventually lead to an incident or an accident.

2. How can the risk of selective reporting of hazards be mitigated? (AM.I.L.R1)

The hazard of selective reporting can be mitigated by emphasizing that all hazards be reported, no matter how insignificant they might seem. This discipline should be encouraged for all types of hazard identification (reactive, proactive, and predictive approaches).

3. What are some of the results of fatigue in an aviation mechanic? (AM.I.L.R2)

The results of fatigue include lack of judgement, inability to concentrate, decreased cognitive ability, reduced reflex time, and lack of coordination.

4. Are there other types of fatigue besides physical fatigue? (AM.I.L.R2)

Fatigue can be either physical or emotional. Emotional stress can be caused by a variety of factors from both within and from outside the workplace.

5. Describe the different parts of an effective fatigue risk management system (FRMS). (AM.I.L.R2)

An effective FRMS requires a systematic approach to safety reporting, integrated company policies, senior management commitment, incident reporting and analysis, continuous monitoring, and proactive risk assessment.

6. **Is a safety management system (SMS) required to have a successful fatigue risk management system?** (AM.I.L.R2)

   No. A fatigue risk management system can be implemented as part of an SMS or as a stand-alone effort to reduce the risk associated with fatigue.

7. **Describe some non-invasive tools that can be used to monitor and mitigate risk.** (AM.I.L.R3)

   A non-invasive tool is one that does not directly involve or interfere with an individual performing a task. Non-invasive tools include safety reporting, integrated company policies, senior management commitment, incident reporting and analysis, continuous monitoring, and proactive risk assessment.

8. **What are some positive actions that an aviation mechanic can take to promote a safe and professional workplace?** (AM.I.L.R3)

   The "Magnificent Seven" attitudes were developed to help aviation mechanics focus on positive actions and attitudes. These are:

   1. We work to accentuate the positive and eliminate the negative.
   2. Safety is not a game because the price of losing is too high.
   3. Just for today—zero error.
   4. We all do our part to prevent Murphy from hitting the jackpot.
   5. Our signature is our word and more precious than gold.
   6. We are all part of the team.
   7. We always work with a safety net.

Skills

1. **Fill out a Form 8010-4 (MFDR) on an assigned aircraft or component.** (AM.I.L.S1)

2. **Demonstrate to the examiner the correct way to complete an FAA Form 8010-4, Malfunction or Defect Report.** (AM.I.L.S1)

3. **Properly complete a Malfunction or Defect Report on a problem that is specified by the examiner.** (AM.I.L.S1)

4. **For general aviation aircraft, operating under Part 91, demonstrate to the examiner where to find the reporting requirements of failures, malfunctions, and defects, as specified in the Code of Federal Regulations.** (AM.I.L.S1)

   Note: Reference 14 CFR §21.3 and §91.1415 (as applicable).

5. **Demonstrate to the examiner where to find the FAA database of processed reports (Malfunction or Defect and Service Difficulty Reports).** (AM.I.L.S1)

   Note: Found under the FAA Service Difficulty Reporting System (SDRS).

6. **Create a written shift turnover log for an aircraft undergoing an engine change.** (AM.I.L.S2)

7. **Present a simulated oral brief for a shift turnover for an aircraft that is in the middle of a progressive inspection. The examiner will take the role of the incoming shift supervisor.** (AM.I.L.S2)

8. **Create a list of questions that you would ask at the beginning of a shift where you are taking over the replacement of a main landing gear on a business jet.** (AM.I.L.S2)

9. **Demonstrate to the examiner where to find AC 120-72.** (AM.I.L.S3)

10. **Demonstrate to the examiner where to find the FAA publication *The Operator's Manual for Human Factors in Maintenance and Ground Operations.*** (AM.I.L.S3)

11. **Demonstrate to the examiner where to find DOT/FAA/AM-11/10, *Fatigue Risk Management in Aviation Maintenance: Current Best Practices and Potential Future Countermeasures.*** (AM.I.L.S3)

The Airframe Oral and Practical Tests

There are 15 subject areas that are tested on the Airframe Oral and Practical Exams.

For each subject area, this guide provides typical oral questions and succinct answers for the knowledge and risk management elements and presents the skills that applicants must understand and demonstrate.

II. Airframe

A. Metallic Structures
B. Non-Metallic Structures
C. Flight Controls
D. Airframe Inspection
E. Landing Gear Systems
F. Hydraulic and Pneumatic Systems
G. Environmental Systems
H. Aircraft Instrument Systems
I. Communication and Navigation Systems
J. Aircraft Fuel Systems
K. Aircraft Electrical Systems
L. Ice and Rain Control Systems
M. Airframe Fire Protection Systems
N. Rotorcraft Fundamentals
O. Water and Waste Systems

A. Metallic Structures

References: AC 43.13-1; FAA-H-8083-31

Knowledge

1. **What are key areas to consider when inspecting a metal structure?** (AM.II.A.K1)
   a. Inspect riveted joints, looking for chipped or cracked paint that may indicate shifted or loose rivets. Smoking around rivets indicates movement, or fretting rivets. Check for cracked rivets and missing shop and manufactured heads.
   b. Look for cracks near rivets or stressed areas.
   c. Inspect for corrosion.

2. **What are the types of damage and defects that may be observed on sheet metal aircraft parts?** (AM.II.A.K2)
   Brinelling, burnishing, burr, corrosion, crack, cut, dent, erosion, chattering, galling, gouge, inclusion, nick, pitting, scratch, score, stain, upsetting.

3. **What is one of the most important considerations when selecting materials for a sheet metal repair?** (AM.II.A.K3)
   The repair material must duplicate the strength of the original structure.

4. **For maximum strength of a formed sheet metal fitting, should the bend be made along or across the grain of the metal?** (AM.II.A.K4)
   Across the grain.

5. **What determines the minimum bend radius that can be used when forming a sheet metal structural fitting?** (AM.II.A.K4)
   The alloy, the metal thickness, and its hardness.

6. **What is meant by the neutral axis in a sheet of metal?** (AM.II.A.K4)
   A plane within the metal that neither stretches nor shrinks when the metal is being bent.

7. **What is a mold line in the development of a flat pattern for a formed metal part?** (AM.II.A.K4)
   An extension of the flat sides beyond the radius.

8. **What is the bend tangent line?** (AM.II.A.K4)
   A line in a flat pattern layout at which the bend starts.

9. **What is meant by setback?** (AM.II.A.K4)
   The distance the jaws of a brake must be set back from the mold line to form a bend.

10. **What is meant by bend allowance?** (AM.II.A.K4)
    The actual amount of metal in a bend. It is the distance between the bend tangent lines in a flat pattern layout.

11. **What is a sight line?** (AM.II.A.K4)
A line drawn on a flat pattern layout within the bend allowance that is one bend radius from the bend tangent line. When the sight line is directly below the nose of the radius bar on the brake, the bend will start at the bend tangent line.

12. **What is the main function of throatless shears in an aircraft sheet metal shop?** (AM.II.A.K4)
Throatless shears are used to cut mild carbon steel up to 10-gauge, and stainless steel up to 12-gauge. They can be used to cut irregular curves in the metal.

13. **What kind of metal forming is done by a slip roll former?** (AM.II.A.K4)
Simple curves with a large radius.

14. **What kind of metal forming is done by bumping?** (AM.II.A.K4)
Compound curves in sheet metal.

15. **When forming a curved angle, what must be done to the flanges?** (AM.II.A.K4)
The flanges must be stretched for a convex curve and shrunk for a concave curve.

16. **When hand-forming a piece of sheet metal that has a concave curve, should the forming be started in the center of the curve, or at its edges?** (AM.II.A.K4)
Start at the edges and work toward the center.

17. **What is meant by a joggle in a piece of sheet metal?** (AM.II.A.K4)
A joggle is a small offset near the edge of a piece of sheet metal that allows the sheet to overlap another piece of metal.

18. **What type of device is a Cleco fastener?** (AM.II.A.K5)
A patented fastener that is inserted in the rivet holes and used to hold two pieces of sheet metal together until they can be riveted.

19. **What type of rivet may be used to replace a round head rivet in an aircraft structure?** (AM.II.A.K5)
A universal head rivet.

20. **How long should a rivet be to join two pieces of sheet metal?** (AM.II.A.K5)
The combined thickness of the metal sheets plus 1-1/2 times the rivet shank diameter.

21. **How can rivets be physically identified for part number and type of material?** (AM.II.A.K5)
The identifying mark on the head of an aluminum alloy rivet indicates the specific alloy used in the manufacture of the rivet.

22. **What type of metal should be hot-dimpled?** (AM.II.A.K6)
7075-T6, 2024-T81 aluminum alloys, and magnesium alloys should be hot-dimpled.

23. **What is the minimum edge distance allowed when installing rivets in a piece of aircraft sheet metal structure?** (AM.II.A.K7)
Two times the diameter of the rivet shank.

24. **What is the recommended transverse pitch to use when making a riveted two row splice in a piece of sheet metal?** (AM.II.A.K7)

    Three-fourths of the distance between the rivets in the rows.

25. **Which stringer attached to a wing skin would require the greatest number of rivets for a splice?** (AM.II.A.K7)

    A stringer in the lower surface.

26. **Why should aluminum alloy rivets be driven with as few blows as is practical?** (AM.II.A.K8)

    Excessive hammering will work-harden the rivets and make them difficult to drive.

27. **What determines whether a piece of sheet metal should be dimpled or countersunk when installing flush rivets?** (AM.II.A.K8)

    The thickness of the sheet. Countersinking should be done only when the thickness of the sheet is greater than the thickness of the rivet head.

28. **What should be done to protect a riveted joint between aluminum alloy and magnesium alloy from corrosion?** (AM.II.A.K8)

    Coat the faying surface with a corrosion-inhibiting primer, dip the rivets in the primer, and drive them while the primer is wet.

29. **What is the purpose of the beehive spring on a rivet gun?** (AM.II.A.K8)

    It holds the rivet set in the gun and allows the gun to vibrate the set.

30. **What must be done to an aircraft fuel tank before it can be repaired by welding?** (AM.II.A.K9)

    The gasoline fumes must all be purged from the tank by running live steam through it for at least 30 minutes, by soaking it in hot water, or by filling it with nitrogen or carbon dioxide.

31. **When drilling out rivets on an airplane or drilling holes for a repair, what precaution should be taken?** (AM.II.A.K9)

    Do not allow the drill to enter into the space behind the structure, as the drill may cause damage to other structures or systems.

32. **What are the two fuel gases most generally used for gas welding?** (AM.II.A.K10)

    Oxygen and acetylene.

33. **What fuel gases are used for welding aluminum?** (AM.II.A.K10)

    Oxygen and hydrogen.

34. **Why is it important that the pressure of the gas in an acetylene cylinder be kept low?** (AM.II.A.K11)

    Acetylene gas becomes unstable when it is kept under pressure of more than about 15 psi.

35. **What are two types of torches used in gas welding?** (AM.II.A.K12)

    Balanced-pressure torches and injector torches.

36. **What determines the amount of heat that is put into a weld by an oxy-acetylene torch?** (AM.II.A.K12)
    The size of the orifice in the torch tip.

37. **What is the difference in the appearance of an oxidizing flame, a neutral flame, and a reducing flame produced by an oxy-acetylene torch?** (AM.II.A.K12)
    An oxidizing flame has a pointed inner cone, and the torch makes a hissing noise. A neutral flame has a rounded inner cone, and there is no feather around the inner cone. A reducing flame has a definite feather around the inner cone.

38. **What is one method of minimizing distortion when making a long butt weld?** (AM.II.A.K12)
    Skip welding minimizes distortion. Tack weld the materials together and then complete the welds between the tacks, starting at a tack and working back toward the finished weld.

39. **What is meant by tack welding?** (AM.II.A.K12)
    Tack welding is the use of small welded spots to hold the material together until the final bead is run.

40. **Why must thick plates of metal be preheated before they are welded?** (AM.II.A.K12)
    Preheating is a method of controlling the expansion and contraction of the metal being welded. Preheating minimizes the stresses that are caused when welding thick metal.

41. **What must be done to an aircraft structure after it has been welded while clamped in a heavy jig or fixture?** (AM.II.A.K12)
    It must be normalized by heating it to a uniform red heat and allowed to cool slowly in still air.

42. **Why is it important that all traces of the welding flux be removed after a piece of aluminum or magnesium is welded?** (AM.II.A.K12)
    Welding flux is corrosive and it must be removed to keep the metal from corroding.

43. **What is the difference between brazing and welding?** (AM.II.A.K12)
    In brazing, the base metal is not melted, but is covered with a low-melting-point alloy. In welding, the base metal is melted.

44. **What kind of flame should be used when gas welding aluminum?** (AM.II.A.K12)
    A soft, neutral oxy-hydrogen flame.

45. **What is an acceptable acetylene line pressure to use when welding with an oxy-acetylene rig?** (AM.II.A.K12)
    About 5 psi.

46. **What kind of flame should be used when gas welding stainless steel?** (AM.II.A.K12)
    A slightly carburizing flame.

47. **How much should the bead penetrate the material when welding two pieces of steel with a butt weld?** (AM.II.A.K12)
    The joint should have 100% penetration.

48. **What is meant by a soft flame?** (AM.II.A.K12)
A soft flame is one that is made when the pressures of the gases are low enough that the flame does not make a noise and does not blow the puddle.

49. **What color lenses are used for gas welding steel?** (AM.II.A.K12)
Green or brown.

50. **What color lenses are used for gas welding aluminum?** (AM.II.A.K12)
Blue.

51. **What is another name for GTA welding?** (AM.II.A.K13)
TIG welding.

52. **What is used as the electrode in GTA welding?** (AM.II.A.K13)
A small-diameter tungsten wire.

53. **What are three types of power that can be used for GTA welding?** (AM.II.A.K13)
DC-straight polarity, DC-reverse polarity, and AC.

54. **Which type of power provides the greatest heat and deepest penetration in GTA welding?** (AM.II.A.K13)
DC-straight polarity.

55. **Which type of power is used for GTA welding aluminum and magnesium?** (AM.II.A.K13)
DC-reverse polarity.

56. **What type of shielding gases are used for GTA welding?** (AM.II.A.K14)
Helium and argon.

57. **What is the function of the inert gas used in GTA and GMA welding?** (AM.II.A.K14)
The inert gas forms a shield to keep oxygen away from the weld puddle so oxides cannot form and weaken the weld.

58. **Why is GTA welding preferred over oxy-acetylene welding for building and repairing welded steel tube aircraft structure?** (AM.II.A.K15)
The heat is concentrated in the weld and does not cause as much distortion as gas welding.

59. **What type of repair can be used when a steel structural tube is dented to more than 1/4 of its circumference?** (AM.II.A.K15)
A patch may be welded over the damage.

60. **What type of repair can be used when a steel structural tube is dented at a cluster?** (AM.II.A.K15)
A patch may be welded over the damaged area with fingers extending up along each member of the cluster.

61. **What is the preferred method of splicing a new piece of tubing into a structure?** (AM.II.A.K16)
An inner sleeve splice.

62. **How is the inner sleeve tube held in the structural tube until the gap is welded?** (AM.II.A.K16)
It is held in place with rosette welds.

63. **What is one limitation of using an outer-sleeve splice in an aircraft structure?** (AM.II.A.K16)
This type of splice is not suitable where it would cause a bulge in the fabric.

64. **How should the ends of an outer-sleeve be cut?** (AM.II.A.K16)
Cut it with a 30° fishmouth.

65. **Should a riveted joint fail in shear or in bearing?** (AM.II.A.K17)
It should fail in shear. The rivets should shear before the sheet tears at the rivet holes.

Risk Management

1. **Explain the risks associated with choosing the incorrect materials for a repair.** (AM.II.A.R1)
The repair material should be the same strength as the original material. Material that is too weak will cause the repair to be weaker than the original. Material that is stronger than the original will cause stresses to move to other structure, increasing the risk of failure to the surrounding structure. Choosing a dissimilar metal for the repair increases the chance of corrosion.

2. **What are some precautions and safety practices that should be used when working with sheet metal repairs?** (AM.II.A.R2)
   - Use caution when working around equipment such as metal shears and breaks which can cause personal injury.
   - Freshly cut sheet metal has sharp edges that can cause personal damage or scratches and gouges on adjoining structure.
   - When drilling out rivets or rivet holes, metal shavings will be present. Use extra caution when drilling overhead.
   - Use caution with drills. Never place fingers or hands behind the area being drilled and be careful to not damage surrounding structure or systems when the drill bit passes through.

3. **What PPE (personal protection equipment) should be used when making a sheet metal repair?** (AM.II.A.R3)
Use eye protection at all times. Use gloves when working with sharp edges or drills. Use ear protection when working in closed areas, especially when driving rivets.

4. **Explain the proper handling and storage of compressed gas bottles.** (AM.II.A.R4)

   When handling compressed gas cylinders:

   - Be careful not to drop or bang cylinders, and prevent cylinders from striking each other or hard surfaces.
   - Secure cylinders in a boot or by chain to a fixed support to prevent them from being dropped or from falling over.
   - Do not use the valve cover (cap) to lift a cylinder.

   When storing cylinders:

   - Always keep valve protection caps on cylinders except when the cylinder is being used or serviced.
   - Store them in an upright position and secure them with a chain, strap, or cable to a stationary building support.
   - Store in a location where they will not be subject to mechanical or physical damage, heat, or electrical circuits.
   - Separate compressed gas cylinders from each other based on the hazard class of their contents.
   - Oxygen cylinders should be kept a minimum of 20 feet away from flammable gases, liquids, solids, greases, and oils.
   - Do not expose cylinders to an open flame, or any temperatures above 125 degrees Fahrenheit.

5. **What precautions should be taken when using electric welding equipment?** (AM.II.A.R5)

   Electric welding requires the completion of a circuit. It is important that this circuit is kept complete, and the materials being welded must be well grounded as well as the surface area or bench where the welding is taking place. Poor electrical grounds can cause electric shock to the individual doing the welding.

Skills

1. **Select the correct length and diameter of rivets furnished by the examiner, and properly install them to join two pieces of aluminum alloy sheet.** (AM.II.A.S1)

2. **Properly remove a series of rivets from a piece of aircraft structure.** (AM.II.A.S1)

3. **Demonstrate to the examiner the correct way to set up and use a squeeze riveter.** (AM.II.A.S1)

4. **Properly install a series of blind fasteners and demonstrate the proper technique for removing a blind fastener.** (AM.II.A.S2)

5. **Evaluate a sheet metal repair and determine the applicability of the repair.** (AM.II.A.S3)

6. **Install different types of special purpose fasteners into a metallic structure.** (AM.II.A.S4)

7. **Design a repair specified by the examiner using the manufacturer's structural repair manual.** (AM.II.A.S5)

8. **Prepare and install a metal patch to repair damage.** (AM.II.A.S6)

9. **Create a drawing for a repair specified by the examiner. Include dimensions, sheet metal thicknesses, and rivet location.** (AM.II.A.S7)

10. **Remove a riveted repair provided by the examiner.** (AM.II.A.S8)

11. **Given a piece of metal with some improperly driven rivets in it, identify the cause of the bad rivets.** (AM.II.A.S8)

12. **Lay out and form a channel of specified dimensions from a piece of aluminum alloy sheet. Use the minimum bend radius allowed for the material.** (AM.II.A.S9)

13. **Trim and form sheet metal to fit into a prepared area.** (AM.II.A.S9)

14. **Fabricate an aluminum part in accordance with a drawing provided by the examiner.** (AM.II.A.S10)

15. **Lay out a rivet pattern for a patch of a size specified by the examiner.** (AM.II.A.S11)

16. **Countersink holes in sheet metal assigned by the examiner with a .010" tolerance.** (AM.II.A.S12)

17. **Perform a repair on a damaged aluminum sheet.** (AM.II.A.S13)

18. **Examine a damaged structure and determine if a repair can be made.** (AM.II.A.S14)

B. Non-Metallic Structures

References: AC 43.13-1; FAA-H-8083-31

Knowledge

1. **Why must abrupt changes in the cross-sectional area of a wooden structural member be avoided?** (AM.II.B.K1)
   Abrupt changes in the cross sectional area of a structural member concentrate stresses and can cause failure.

2. **How can you determine the slope of the grain in a piece of wood to be used in a wing spar?** (AM.II.B.K1)
   Suitable wood for aircraft must have a slope that is not steeper than 1:15.

3. **How much pressure must be applied to a glue joint in a piece of softwood to produce a strong joint?** (AM.II.B.K1)
   125 to 150 pounds per square inch.

4. **Why should sandpaper never be used when preparing a scarf joint in a wing spar for splicing?** (AM.II.B.K1)
   The dust caused by sanding will plug the pores of the wood so the glue cannot get in to form a good bond.

5. **Why are light steel bushings often used in bolt holes in a wood wing spar?** (AM.II.B.K1)
   The bushing keeps the spar from being crushed when the nut on the attachment bolt is tightened.

6. **Why should wooden wing spars be finished with a transparent varnish?** (AM.II.B.K1)
   The transparent finish allows any decay or rot that develops in the wood to be detected.

7. **What effect does moisture have on a wood aircraft structure?** (AM.II.B.K2)
   Moisture causes the wood to swell and crack as it dries out. It allows fungus to develop in the wood and cause it to decay.

8. **Which species of wood is considered to be the standard when comparing other woods for use in aircraft structure?** (AM.II.B.K3)
   Sitka spruce.

9. **What is the basic difference between laminated wood and plywood?** (AM.II.B.K3)
   In laminated wood, all of the grain runs in the same direction. In plywood, the grain in the layers cross the others at a 90° or 45° angle.

10. **Which wood is more inclined to warp, flat grain or vertical grain?** (AM.II.B.K3)
    Flat grain.

11. **What are three types of wood used in aircraft structures?** (AM.II.B.K3)
    Spruce, Douglas fir, noble fir, western hemlock, northern white pine, Port Orford white cedar, yellow poplar.

12. **What is the basic wood used for aircraft wing spars?** (AM.II.B.K3)
    Sitka spruce.

13. **Can northern white pine be used as a substitute for spruce?** (AM.II.B.K4)
    Yes, but it must be increased in size to compensate for its lower strength.

14. **How do you detect decay in a wood structure?** (AM.II.B.K5)
    Stick a sharp-pointed knife blade in the suspect area and pry the wood up. If the wood is good, it will come up as a long splinter; if it is decayed it will come up as a chunk.

15. **Are mineral streaks in a piece of structural aircraft wood reason for rejecting the wood?** (AM.II.B.K5)
    No, if there is no evidence of decay in the wood.

16. **How is compression wood identified?** (AM.II.B.K5)
    It has a high specific gravity, it appears to have an excessive growth of summerwood, and it has little contrast between springwood and summerwood.

17. **Is a hard knot that is 1/2-inch in diameter allowed if it is in the web of a wing spar?** (AM.II.B.K5)
No, 3/8-inch diameter is the maximum allowable knot, and it must meet severe restrictions.

18. **Are pin knot clusters allowable in aircraft structural wood?** (AM.II.B.K5)
Yes, if they cause only a small effect on grain direction.

19. **What kind of glue is recommended for making repairs to a wood aircraft structure?** (AM.II.B.K6)
Synthetic resin or resorcinol glue.

20. **What reference material may be used for acceptable repairs to wood aircraft structure?** (AM.II.B.K6)
AC 43.13-1B, Chapter 1.

21. **How is a scarf splice on a wing spar reinforced?** (AM.II.B.K6)
Solid spruce or birch plywood reinforcing plates are glued to each side of the spar, centered at each end of the scarf.

22. **What is the correct repair to a wooden aircraft wing spar if the wing-attach bolt holes in the spar are elongated?** (AM.II.B.K6)
Splice in a new section of the spar and drill new holes.

23. **What is the minimum taper to use when repairing a wood wing rib cap strip?** (AM.II.B.K6)
12 times the thickness recommended; 10 times is minimum.

24. **How is aircraft plywood prepared for making a compound bend?** (AM.II.B.K6)
Soak the wood in hot water until it is pliable.

25. **What is used to apply pressure to a glued joint when splicing a wood aircraft wing spar?** (AM.II.B.K6)
Cabinetmakers' parallel clamps.

26. **What kind of repair is recommended for a hole in the plywood skin of an aircraft wing?** (AM.II.B.K6)
A scarf patch.

27. **What is the recommended taper for a splayed patch in a plywood aircraft skin?** (AM.II.B.K6)
5 to 1.

28. **What is the recommended taper for a scarf patch in a plywood aircraft skin?** (AM.II.B.K6)
12 to 1.

29. **What is the largest hole in a plywood wing skin that can be repaired with a fabric patch?** (AM.II.B.K6)
1-inch in diameter.

30. **How long should a glue joint be kept under pressure when splicing a wood aircraft wing spar?** (AM.II.B.K6)
For at least 7 hours.

31. **Which area of a wood wing spar must not contain any splice?** (AM.II.B.K6)
There must be no splice under wing attach fittings, landing gear fittings, engine mount fittings, or lift and interplane strut fittings.

32. **What is done to a splice in a wood aircraft wing spar to strengthen the splice?** (AM.II.B.K6)
Reinforcing plates are glued to both sides of the splice.

33. **When an aircraft is being recovered, when is fungicidal dope applied to the fabric?** (AM.II.B.K6)
With the first coat of dope that is brushed into the fabric.

34. **What is the recommended type of repair to a fabric-covered aircraft surface when it has an L-shaped tear, with each of the legs of the tear more than 14 inches long?** (AM.II.B.K7)
If the never-exceed speed of the aircraft is less than 150 miles per hour, a doped-on repair can be made.

35. **How is aircraft performance related to the determination of appropriate fabric covering?** (AM.II.B.K7)
Aircraft wing loading and certified never exceed speeds (V_{NE}) determine the fabric strength requirements.

36. **What reference material may be used for acceptable covering methods for fabric-covered aircraft structure?** (AM.II.B.K8)
AC 43.13-1B, Chapter 2.

37. **What are three types of fabric that can be used to cover an aircraft?** (AM.II.B.K8)
Cotton fabric, polyester fabric, and glass fabric.

38. **What material is used for inter-rib bracing in a fabric-covered aircraft wing?** (AM.II.B.K8)
Cotton reinforcing tape.

39. **What are the two basic types of dope used on fabric-covered aircraft?** (AM.II.B.K8)
Nitrate dope and butyrate (cellulose acetate butyrate, or CAB) dope.

40. **What kind of dope is used on polyester synthetic fabric that has been heat-shrunk on an aircraft structure?** (AM.II.B.K8)
Non-tautening butyrate dope.

41. **What type of fabric is most widely used for covering aircraft structures?** (AM.II.B.K8)
Heat-shrinkable polyester fabric.

42. What is the preferred seam used for machine-sewing pieces of aircraft fabric together? (AM.II.B.K9)

The French fell seam.

43. Should a sewed seam in the fabric used to cover an aircraft wing run spanwise or chordwise? (AM.II.B.K9)

Both spanwise and chordwise seams are permissible, but chordwise seams are preferred.

44. What type of hand-sewing stitch is used when sewing in a panel of new fabric on an aircraft fabric-covered wing? (AM.II.B.K9)

A baseball stitch, locked every eight to ten stitches.

45. What is meant by the selvage edge of a piece of fabric? (AM.II.B.K10)

It is the woven edge of fabric used to prevent the material unraveling during normal handling.

46. What is an antitear strip, and when are they required on a fabric-covered aircraft? (AM.II.B.K10)

An antitear strip is a strip of the same type of fabric as is used for covering the wings. It is laid over the rib between the reinforcing tape and the fabric. An antitear strip is required for aircraft that have a never-exceed speed in excess of 250 miles per hour.

47. Why are some portions of the structure of an aircraft dope proofed before they are covered with fabric? (AM.II.B.K11)

Dope proofing keeps the fabric from sticking to the structure when the first coat of dope is applied. The fabric normally sags enough to touch the structure before it begins to pull taut.

48. What are two methods of applying the fabric to the wing of an airplane? (AM.II.B.K12)

The blanket method and the envelope method.

49. What material is used to cover the overlapping edges of the leading edge metal to protect the fabric? (AM.II.B.K12)

Cloth tape.

50. How are the wrinkles pulled out of cotton fabric? (AM.II.B.K12)

Wet the cotton with water and allow it to dry.

51. How is polyester fabric shrunk on the aircraft structure? (AM.II.B.K12)

With heat from an iron.

52. Where are drainage grommets located on a fabric-covered aircraft wing? (AM.II.B.K12)

At the lowest point in each bay. It is customary to install a grommet on each side of a wing rib, on the underside of the wing, at the trailing edge.

53. How wide should the surface tape be that is used to cover the trailing edge of an aircraft wing? (AM.II.B.K12)

Three inches wide.

54. **What type of rib lacing cord is recommended for attaching cotton fabric to an aircraft structure?** (AM.II.B.K13)
Waxed linen cord.

55. **What are two methods of attaching the fabric to the aircraft structure?** (AM.II.B.K13)
By sewing and with a cement.

56. **What type of knot is used for locking the stitches that are used for rib lacing on a fabric-covered aircraft wing?** (AM.II.B.K13)
A modified seine knot.

57. **What determines the spacing of the rib lacing stitches on a fabric-covered aircraft wing?** (AM.II.B.K13)
The never-exceed speed of the aircraft.

58. **Why is the surface tape used on the trailing edge of control surfaces of some airplanes notched?** (AM.II.B.K14)
Since the edges of this tape face into the wind, it is possible that it could start to lift and form a very effective spoiler. If the tape is notched, it will tear off at a notch.

59. **What is the purpose of the reinforcing tape used between the fabric and the rib lacing on an aircraft wing?** (AM.II.B.K14)
The reinforcing tape keeps the rib lacing cord from pulling through the fabric.

60. **What areas of a fabric-covered aircraft are more susceptible to deterioration?** (AM.II.B.K14)
All areas exposed to the elements, and especially those exposed to ultraviolet (UV) radiation, such as the upper surfaces of the aircraft.

61. **What is done to cotton and linen fabric to protect it from mildew?** (AM.II.B.K15)
The first coat of dope used on cotton and linen fabric has a mildewcide mixed in it.

62. **Does rejuvenator restore strength to the fabric?** (AM.II.B.K15)
No, it only restores resilience to the finish.

63. **What is the minimum strength to which aircraft fabric is allowed to deteriorate before it is considered to be unairworthy?** (AM.II.B.K16)
Fabric can deteriorate to 70% of the strength of the fabric required for the aircraft.

64. **How is the strength of the fabric on an aircraft structure determined?** (AM.II.B.K16)
An approximate strength test can be made with an FAA-approved fabric punch tester, but the only way to know for sure that the fabric has sufficient strength is by pull-testing a one-inch-wide sample of the fabric.

65. **What paperwork must be completed if an aircraft that was originally covered with Grade-A cotton fabric is re-covered using a synthetic fabric?** (AM.II.B.K17)
The covering must be done according to a Supplemental Type Certificate, and a Form 337 must be executed, stating that all materials and processes complied with the requirements of the STC.

66. When is the finishing tape applied to a fabric-covered wing when it is being recovered? (AM.II.B.K17)

After the second coat of dope has dried and the nap of the fabric has been sanded off.

67. When are drainage grommets applied when an aircraft is being re-covered? (AM.II.B.K17)

They are laid into the third coat of dope, at the same time the surface tape is applied.

68. What will happen if dope is sprayed over an enameled surface? (AM.II.B.K17)

The thinner in the dope will penetrate the enamel surface and cause it to swell.

69. What can be done to remedy blushing that has formed on a doped surface that has just been sprayed? (AM.II.B.K17)

Spray a very light mist coat of a mixture of one part retarder to two parts of thinner over the blushed area. Allow it to dry and spray on another coat. If this does not remove the blush, the blushed dope will have to be sanded off and new dope applied.

70. What is a rejuvenator? (AM.II.B.K17)

A slow-drying finishing material that has potent solvents and plasticizers that soften a dried surface film and restore resilience to the film.

71. Why is the first coat of dope brushed onto the fabric? (AM.II.B.K17)

To ensure thorough penetration and encapsulation of all of the fibers.

72. What is the general reason for runs and sags in a finish that is being sprayed onto a flat surface? (AM.II.B.K17)

Too much dope is being applied. The film is too thick.

73. What is the most common cause for dope roping? (AM.II.B.K17)

The dope was improperly thinned or it was too cold.

74. What causes pinholes in a dope finish? (AM.II.B.K17)

Excessive atomizing air pressure on the spray gun.

75. Why should the first coat of dope be thinned? (AM.II.B.K17)

To ensure thorough penetration and encapsulation of all of the fibers.

76. Why is aluminum pigmented dope used on a fabric-covered aircraft? (AM.II.B.K17)

To protect the clear dope and the fabric from the harmful effects of the sun.

77. Why should a fabric-covered surface be electrically grounded when dry-sanding it? (AM.II.B.K17)

Dry-sanding can create enough static electricity on the surface that it can cause a spark and ignite the dope fumes inside the structure.

78. What instructions must be followed when covering and finishing an aircraft with materials specified in a Supplemental Type Certificate (STC)? (AM.II.B.K17)

The instructions that are a part of the STC.

79. **What finishing materials must be used when recovering an aircraft using an STC?** (AM.II.B.K17)
The finishing materials specified by the STC.

80. **How can water entrapped in a honeycomb structure be detected?** (AM.II.B.K18)
By the use of radiographic inspection.

81. **What effect can entrapped moisture have on metal honeycomb structure?** (AM.II.B.K18)
Entrapped water can cause corrosion.

82. **What type of nondestructive inspection is suitable for detecting internal damage in honeycomb material?** (AM.II.B.K18)
Ultrasonic inspection.

83. **What is the simplest way to detect delamination in a composite structure?** (AM.II.B.K18)
Tap the surface with the edge of a coin. If the material is sound, the tapping will result in a clear ringing sound; but if it is delaminated, the sound will be a dull thud.

84. **What are the types of defects that can be found in composite structures?** (AM.II.B.K19)
Manufacturing defects include delamination, resin-starved areas, resin-rich areas, blisters, air bubbles, wrinkles, voids, and thermal decomposition.

    In-service defects include environmental degradation, impact damage, fatigue, cracks from local overload, debonding, delamination, fiber fracturing, and erosion.

85. **What are two advantages of laminated construction over riveted sheet metal?** (AM.II.B.K20)
Light weight and rigidity.

86. **What are two popular types of core material used in laminated structure?** (AM.II.B.K20)
Foam and honeycomb.

87. **What are two popular types of matrix material used in laminated structure?** (AM.II.B.K20)
Polyester and epoxy resins.

88. **What are the two parts of a polyester matrix material?** (AM.II.B.K20)
Resin and catalyst.

89. **What are the two parts of an epoxy matrix material?** (AM.II.B.K20)
Resin and hardener.

90. **What are three materials that may be used to reinforce the matrix material for aircraft structure?** (AM.II.B.K20)
Fiberglass, Kevlar®, and graphite.

91. **What is meant by a unidirectional fabric?** (AM.II.B.K20)
A fabric in which all of the major fibers run in the same direction.

92. **What is meant by the ribbon direction of a honeycomb material?** (AM.II.B.K20)
The direction in a piece of honeycomb material that is parallel to the length of the strips of material that make up the core.

93. **What is meant by the shelf life of a material?** (AM.II.B.K21)
The normal length of time a material can be expected to keep its usable characteristics if it is stored and not used.

94. **What is meant by the pot life of a resin?** (AM.II.B.K21)
The length of time a resin will remain workable after the catalyst has been added.

95. **Where are pre-impregnated (prepreg) materials normally stored?** (AM.II.B.K21)
In a freezer, at temperatures specified by the manufacturer.

96. **What kind of repair can be made to a small damage of the core material and one face sheet of a piece of aluminum alloy honeycomb structure?** (AM.II.B.K22)
A potted compound repair.

97. **How do you grind the point of a twist drill that is to be used for drilling transparent acrylic material?** (AM.II.B.K22)
The cutting edge should be dubbed off to a zero rake angle, and the included angle of the tip should be ground to 140 degrees.

98. **What is a warp clock in a structural repair manual?** (AM.II.B.K22)
An alignment indicator to show the orientation of the plies of a composite material. The ply direction is shown in relation to a reference direction.

99. **What are two types of repair to a damaged honeycomb core composite material?** (AM.II.B.K22)
Room-temperature cure repair and hot-bond repair.

100. **Of what class of resins are aircraft windows and windshields made?** (AM.II.B.K23)
Thermoplastic.

101. **What is meant by crazing of a transparent plastic material?** (AM.II.B.K23)
Tiny hair-like cracks that may not extend all of the way to the surface. Crazing is caused by stresses or chemical fumes.

102. **How should sheets of transparent plastic material be stored?** (AM.II.B.K24)
Leave the protective paper on the material and store it tilted approximately 10° from the vertical.

103. **What are the two types of resins used in aircraft construction?** (AM.II.B.K25)
Thermoplastic and thermosetting.

104. **What is necessary to cure a thermosetting resin?** (AM.II.B.K25)
Heat.

105. How tight should the screws be tightened when installing an acrylic windshield in a channel? (AM.II.B.K25)

Tighten the screw to a firm fit and back it off one full turn.

106. What type of material is used to remove surface damage to a piece of acrylic resin? (AM.II.B.K26)

Micro-Mesh® abrasive sheets.

107. What type of transparent plastic material is used for most aircraft windshields? (AM.II.B.K26)

Acrylic plastic.

108. What are the proper methods of repairing windows? (AM.II.B.K27)

Repairs to windows are to be made only on unpressurized aircraft.

- A temporary repair can be made by first stop-drilling the crack, drilling small holes along the crack, and lacing with brass safety wire. Another method of a temporary repair is to stop drill the cracks and clamp the crack together with machine screws, nuts, and washers. In both cases, fill the cracks with clear silicone to waterproof the repair.
- In windshields or side windows with small cracks that are cosmetic only and do not affect the airworthiness, repairs can be made by stop-drilling the ends of the cracks and then using a hypodermic syringe and needle to fill the crack with polymerizable cement such as PS-30 or Weld-On 40.

109. What special precautions must be taken when repairing a radome? (AM.II.B.K28)

Nothing must be done to the radome that will affect its electrical transparency or its aerodynamic strength.

110. What document lists the safety hazards associated with a resin used in composite structure? (AM.II.B.K28)

The Material Safety Data Sheet for that resin.

111. What are the critical areas of inspection on seat belts, shoulder harnesses, and aircraft upholstery? (AM.II.B.K29)

- Verify that seat belts are manufactured in accordance with TSO C22.
- Inspect belt webbing for fraying and tears.
- Inspect buckles and inserts for condition and wear.
- Inspect upholstery for condition and wear. Replace upholstery only with fire-resistant materials approved for aircraft use.

Risk Management

1. Why is it important to select the appropriate glue or fastener for aircraft structures? (AM.II.B.R1)

The use of any glue or fastener not specified by the aircraft manufacturer may render the aircraft unairworthy and unsafe.

2. **Explain the risks associated with composite structural repairs and how to mitigate the risks.** (AM.II.B.R2)
   It is critical that structural repairs are accomplished in accordance with the aircraft manufacturer's instructions or with FAA guidance. Repairs that are not made in accordance with approved data may be unairworthy and unsafe.

3. **What personal protection equipment (PPE) should be used when performing a composite repair?** (AM.II.B.R3)
   Eye protection, breathing protection, and protection of skin and clothing. If power equipment is in use, use hearing protection.

4. **What factors should be considered in the storage of composite materials?** (AM.II.B.R4)
   Most composite materials have shelf-life limitations that are dependent on proper climate control. Some materials, such as prepregs, should be kept in a freezer at temperatures specified by the manufacturer. Any time spent outside of the freezer should be tracked and recorded in accordance with the material specifications.

5. **What are the possible results of improper measuring or mixing of materials used with composite construction?** (AM.II.B.R5)
   A resin-rich application can add unnecessary weight. A resin-starved application will show the fibers through the surface. Improper measuring or mixing of materials can compromise the bond and strength.

6. **Why is it important to only use materials that are approved for a composite repair?** (AM.II.B.R6)
   It is critical that composite repairs are accomplished in accordance with the aircraft manufacturer's instructions or with FAA guidance. Repairs that are not made in accordance with approved data may be unairworthy and unsafe.

7. **What are the risks of using materials that have an expired shelf-life?** (AM.II.B.R7)
   The material properties of the expired materials will begin to degrade and strength and adhesion qualities may suffer.

Skills

1. **Determine the appropriate fasteners to use on a given composite structure.** (AM.II.B.S1)

2. **Inspect a damaged fiberglass component provided by the examiner and perform the appropriate repair.** (AM.II.B.S2)

3. **Explain to the examiner the type of reinforcing fibers that are used when stiffness is the prime consideration.** (AM.II.B.S2)

4. **Explain to the examiner the importance of ply orientation when laying up a laminated composite component.** (AM.II.B.S2)

5. **Demonstrate to the examiner the correct way to inspect a piece of composite material for delamination.** (AM.II.B.S3)

6. **Demonstrate to the examiner the way to evaluate the surface condition of composite structure.** (AM.II.B.S3)

7. **Explain to the examiner the correct way to use ultrasonic testing equipment to check the integrity of composite structure.** (AM.II.B.S3)

8. **Inspect a composite material structure.** (AM.II.B.S3)

9. **Examine a fabric-covered control surface to determine if the fabric is airworthy.** (AM.II.B.S3)

10. **Demonstrate the correct way to clean an acrylic plastic windshield.** (AM.II.B.S4)

11. **Demonstrate the correct way to remove minor scratches from the surface of a piece of acrylic plastic material.** (AM.II.B.S4)

12. **Inspect acrylic windshields.** (AM.II.B.S4)

13. **Demonstrate the correct way to repair a crack in an aircraft side window.** (AM.II.B.S5)

14. **Identify window enclosure materials.** (AM.II.B.S5)

15. **Demonstrate to the examiner where to find the instructions for tying a modified seine knot and explain the procedure.** (AM.II.B.S6)

16. **Mix dope and the correct thinner to get the proper viscosity for spraying. Demonstrate to the examiner the correct way to spray the dope on an aircraft surface.** (AM.II.B.S7)

17. **Demonstrate how to prepare a composite surface for painting using the component supplied by the examiner.** (AM.II.B.S7)

18. **Demonstrate to the examiner the correct way to check a bonded honeycomb structure for indication of internal delamination.** (AM.II.B.S8)

19. **Demonstrate where to find the standard repair dimensions for the repair provided by the examiner.** (AM.II.B.S9)

20. **Explain to the examiner the correct way to repair a wing spar that has an elongated bolt hole in its root end.** (AM.II.B.S10)

21. **Inspect the extent of damage on the component provided by the examiner and determine if the component is repairable.** (AM.II.B.S11)

22. **Determine whether installing a Supplemental Type Certificate (STC) covering is applicable to a given aircraft.** (AM.II.B.S11)

23. **Repair a small damage to a piece of honeycomb structure.** (AM.II.B.S12)

24. **Install special fasteners into a composite structure.** (AM.II.B.S12)

C. Flight Controls

References: AC 43.13-1; FAA-H-8083-31

Knowledge

1. **What is the purpose of an aileron balance cable?** (AM.II.C.K1)
   It ties the ailerons together in such a way that when one aileron deflects downward, the other one is pulled upward.

2. **What is the most common construction configuration of flexible control cables?** (AM.II.C.K1)
   7 by 19.

3. **What materials are used to manufacture flexible control cables?** (AM.II.C.K1)
   Carbon steel coated with tin or zinc, or corrosion-resistant steel.

4. **What is the definition of a critical fatigue area of a control cable?** (AM.II.C.K2)
   Any working length of a cable where the cable runs over, under, or around a pully, sleeve, or through a fairlead; or any section where the cable is flexed, rubbed, or worked in any manner; or any point within 1 foot of a swaged-on fitting.

5. **How are the ends of two control cables normally connected together?** (AM.II.C.K3)
   With swaged, threaded terminals on the ends of each cable, connected with a turnbuckle.

6. **How much is a fairlead allowed to deflect a control cable?** (AM.II.C.K4)
   No more than 3°.

7. **Why are the control cables of large airplanes normally equipped with automatic tension regulators?** (AM.II.C.K4)
   The large amount of aluminum in the aircraft structure contracts so much as its temperature drops in flight that the control cables could become dangerously loose. The automatic tension regulators keep the cable tension constant as the dimensions of the aircraft change.

8. **When rigging control cables, there are stops at the control surfaces that limit travel and stops at the control input. Which stops should the system contact first?** (AM.II.C.K5)
   The stops at the control surface.

9. **How are length adjustments made on a flight control push-pull rod?** (AM.II.C.K6)
   One or both ends of the push-pull rod have threaded rod ends.

10. **What is the purpose of a torque tube in a flight control system?** (AM.II.C.K7)
    To convert a linear motion to an angular or twisting motion.

11. **What component in a flight control system is often used to change motion by 90 degrees?** (AM.II.C.K8)
    A bellcrank.

12. **What is the function of the rudder on an airplane?** (AM.II.C.K9)
    The rudder rotates the airplane about its vertical axis.

13. **What is an aerodynamically balanced control surface?** (AM.II.C.K9)
    A surface with part of its area ahead of the hinge line. When the surface is deflected, the portion ahead of the hinge aids the movement.

14. **In what publication could you find correct control surface movement for a particular airplane?** (AM.II.C.K9)
    In the Type Certificate Data Sheet for the airplane.

15. **Why is it important that control surfaces be statically balanced?** (AM.II.C.K9)
    An out-of-balance control surface can cause severe flutter.

16. **Where can you find the specifications for balancing the control surfaces of an airplane?** (AM.II.C.K9)
    In the aircraft maintenance manual.

17. **What in addition to static unbalance can cause a control surface to flutter?** (AM.II.C.K9)
    Worn hinges or improperly adjusted control cable tension.

18. **What is the function of the elevators on an airplane?** (AM.II.C.K10)
    Elevators rotate the airplane about its lateral axis.

19. **What is a stabilator?** (AM.II.C.K11)
    A single-piece horizontal tail surface that acts as both the horizontal stabilizer and the elevators. A stabilator pivots about its front spar.

20. **What is meant by differential aileron travel?** (AM.II.C.K11)
    Aileron movement in which the upward-moving aileron deflects a greater distance than the one moving downward. The up aileron produces parasite drag to counteract the induced drag produced by the down aileron.

21. **What is a Frise aileron?** (AM.II.C.K11)

    An aileron with its hinge line set back from the leading edge so that when it is deflected upward, part of the leading edge projects below the wing and produces parasite drag to help overcome adverse yaw.

22. **What is a ruddervator?** (AM.II.C.K11)

    Movable control surfaces on a V-tail airplane that are controlled by both the rudder pedals and the control yoke. When the yoke is moved in or out, the ruddervators move together and act as the elevators. When the rudder pedals are depressed, the ruddervators move differentially and act as a rudder.

23. **What is the function of the ailerons on an airplane?** (AM.II.C.K11)

    Ailerons rotate the airplane about its longitudinal axis.

24. **What is the purpose of a spoiler?** (AM.II.C.K12)

    There are spoilers intended for use in flight and others that are used only on the ground. Inflight spoilers are used to aid in roll control. Ground spoilers are used after touchdown to decrease lift and increase drag.

25. **What is the primary purpose of a trim tab?** (AM.II.C.K13)

    To reduce the force needed to move a primary control surface.

26. **Why are flaps used on most aircraft?** (AM.II.C.K13)

    They allow control at lower airspeeds and reduce the speeds needed for takeoff and landing, shortening runway length requirements.

Risk Management

1. **What are the risks associated with misinterpretation of a cable tension chart?** (AM.II.C.R1)

   Not understanding a cable tension chart can lead to incorrect tensions. Cable tensions that are set too high can cause damage to the aircraft and premature wear of the system. Tensions set too low can cause insufficient control travel, leading to loss of aircraft control.

2. **Why is it important to understand the proper procedures for rigging aircraft flight controls?** (AM.II.C.R2)

   Incorrectly rigged flight controls can cause poor aircraft performance, or in the worst case, complete loss of aircraft control.

3. **Where are the proper procedures found for lifting and moving aircraft components into place for installation and assembly?** (AM.II.C.R3)

   In the aircraft maintenance manual.

4. **What risks are associated with the use of incorrect tools and procedures when installing aircraft components and major assemblies?** (AM.II.C.R3)

   Incorrect tools and procedures can cause damage to the aircraft or assemblies, leading to their failure.

5. **Why are cable tensiometers and other rigging equipment maintained under a calibration schedule?** (AM.II.C.R4)

   Proper tensions and accurate dimensions of rigging equipment is critical for airworthiness and aircraft safety.

6. **What risks can arise from the multiple scales that are on some cable tensiometers?** (AM.II.C.R5)

   The use of the wrong scale or adjustment block used on some tools can lead to incorrect cable tensions.

Skills

1. **Explain to the examiner where to find the rigging instruction for the specified control system, and locate the key adjustment location on the aircraft.** (AM.II.C.S1)

2. **Explain to the examiner the operation of a stabilator.** (AM.II.C.S2)

3. **Explain to the examiner the purpose of spoilers on a jet transport airplane.** (AM.II.C.S2)

4. **Explain to the examiner the location and function of all primary and secondary flight controls on the specified aircraft.** (AM.II.C.S2)

5. **Check the flight controls of an airplane, including all of the secondary controls for the correct direction of movement when the flight deck controls are moved.** (AM.II.C.S3)

6. **Locate flight control neutral positions.** (AM.II.C.S3)

7. **Verify the alignment of an empennage or landing gear.** (AM.II.C.S3)

8. **Inspect a primary and secondary control surface assigned by the examiner.** (AM.II.C.S3)

9. **Demonstrate to the examiner the correct way to check the angular deflection of the ailerons on an airplane.** (AM.II.C.S3)

10. **Demonstrate to the examiner the correct way to check the angular movement of the elevators on an airplane.** (AM.II.C.S3)

11. **Demonstrate to the examiner the correct way to check the condition and movement of the trim tabs on an airplane.** (AM.II.C.S3)

12. **Perform an operational check of the flaps and explain to the examiner the way to adjust the sensor to accurately indicate the position of the flaps.** (AM.II.C.S3)

13. **Assemble aircraft components.** (AM.II.C.S3)

14. **Install a control surface on an airplane specified by the examiner. Rig the cables to the proper tension and adjust the stops so the surface will have the correct travel as specified in the TCDS. Properly safety the turnbuckles.** (AM.II.C.S4)

15. **Remove a control surface from an aircraft, check its static balance as instructed in the aircraft maintenance manual, and reinstall it. Check the condition of the hinges and rig the control cables for the proper tension.** (AM.II.C.S4)

16. **Demonstrate to the examiner the correct way to inspect a piece of aircraft control cable for indication of internal corrosion.** (AM.II.C.S5)

17. **Using a control cable tension chart specified by the examiner, find the proper tension for a 3/16, 7 x 19 cable when the aircraft temperature is 80°F.** (AM.II.C.S5)

18. **Inspect a primary control cable system on an aircraft supplied by the examiner.** (AM.II.C.S5)

19. **Locate leveling methods.** (AM.II.C.S6)

20. **Remove and reinstall a primary flight control cable.** (AM.II.C.S6)

21. **Adjust a push-pull control system.** (AM.II.C.S7)

22. **Demonstrate to the examiner the correct way to check the balance of a flight control surface.** (AM.II.C.S8)

23. **Inspect a control system that includes flight control bearings and determine the allowable axial play limits.** (AM.II.C.S9)

24. **Demonstrate to the examiner how to inspect a trim tab system for free play, travel, and operation.** (AM.II.C.S10)

25. **Balance a control surface provided by the examiner.** (AM.II.C.S11)

26. **Properly install a swaged control cable terminal or a Nicopress sleeve type terminal. Explain the way to check the terminal for proper installation.** (AM.II.C.S12)

27. **Demonstrate to the examiner where to find the control travel limits for the specified aircraft.** (AM.II.C.S13)

D. Airframe Inspection

References: AC 43.13-1; FAA-H-8083-31

Knowledge

1. **If a mechanic gives a 100-hour inspection to an aircraft that proves to be unairworthy, what must the mechanic give the owner or operator of the aircraft?** (AM.II.D.K1)

   A signed and dated list of all of the discrepancies that keep the aircraft from being airworthy.

2. **Where can you find the recommended statement to use for recording the approval or disapproval of an aircraft for return to service after a 100-hour inspection?** (AM.II.D.K1)

   In 14 CFR §43.11.

3. **Under what conditions can an aircraft be operated with a 100-hour inspection overdue?** (AM.II.D.K1)

   The aircraft can be operated for no more than 10 hours after an inspection is due for the purpose of flying it to a place where the inspection can be performed.

4. **For how long can an aircraft be operated if a 100-hour inspection is overdue?** (AM.II.D.K1)

   For no more than 10 hours. The time beyond the 100 hours must be subtracted from the time before the next inspection is due.

5. **Under what conditions can an aircraft that is due for an annual inspection be operated?** (AM.II.D.K1)

   It can only be flown when a special flight permit is issued.

6. **What certification is required for a mechanic to be able to approve an aircraft for return to service after a 100-hour inspection?** (AM.II.D.K1)

   An Aviation Mechanic certificate with Airframe and Powerplant ratings.

7. **What determines whether or not an aircraft must be given a 100-hour inspection?** (AM.II.D.K1)

   Aircraft that carry persons for hire and aircraft that are used for flight instruction for hire must be given 100-hour inspections.

8. **What is the difference between an annual inspection and a 100-hour inspection?** (AM.II.D.K1)

   The inspections themselves are identical. An annual inspection can be conducted only by an A&P mechanic who holds an Inspection Authorization, while a 100-hour inspection can be performed by an A&P mechanic without an IA.

9. **What certification is required for a mechanic to conduct an annual inspection and approve the aircraft for return to service after the inspection?** (AM.II.D.K1)
   An Aviation Mechanic certificate with Airframe and Powerplant ratings and an Inspection Authorization.

10. **What is a progressive inspection?** (AM.II.D.K1)
    An inspection of the same level as an annual inspection but approved by the FAA to be performed on a schedule that does not require the aircraft to be out of service for the time necessary to perform the entire inspection all at once.

11. **What certification is required for a mechanic to conduct a progressive inspection?** (AM.II.D.K1)
    An Aviation Mechanic certificate with Airframe and Powerplant ratings and an Inspection Authorization.

12. **Where can you find the requirements for inspecting the altimeter and static systems of aircraft operated under instrument flight rules?** (AM.II.D.K1)
    In 14 CFR Part 43, Appendix E.

13. **Where can you find the requirements for inspecting the ATC transponder that is installed in an aircraft?** (AM.II.D.K1)
    In 14 CFR Part 43, Appendix F.

14. **What items must be inspected on a helicopter in accordance with the instructions for Continued Airworthiness?** (AM.II.D.K1)
    The drive shafts or similar systems, the main rotor transmission gear box, the main rotor and center section, and the auxiliary rotor.

15. **Does the FAA require that a checklist be used when conducting an annual or a 100-hour inspection?** (AM.II.D.K2)
    Yes, according to 14 CFR §43.15(c)(1).

16. **For how long must the record of a 100-hour inspection be retained in the aircraft maintenance records?** (AM.II.D.K2)
    For one year, or until the next 100-hour inspection is completed.

17. **For how long must the record of the current status of life-limited parts of an engine be retained?** (AM.II.D.K2)
    This is part of the permanent records and it must be retained and transferred with the aircraft when it is sold.

18. **Where are the requirements detailed to determine if an AD applies to a specific aircraft?** (AM.II.D.K3)
    In the AD itself.

19. **Who is responsible to ensure that all ADs are complied with?** (AM.II.D.K3)
    The owner or operator of the aircraft.

20. **Where is the current status of life-limited parts found?** (AM.II.D.K4)
    In the aircraft maintenance records.

21. **Who is responsible for maintaining a record of life-limited parts?** (AM.II.D.K4)
The registered owner or operator of the aircraft.

22. **Where would an aviation mechanic find required special inspections?** (AM.II.D.K5)
In the aircraft maintenance manual.

23. **What type of repair requires the use of FAA-approved data?** (AM.II.D.K6)
A major repair.

24. **What type of modification requires the use of FAA-approved data?** (AM.II.D.K6)
A major modification.

25. **Are aircraft maintenance manuals approved by the FAA?** (AM.II.D.K6)
Only Chapter 4, Airworthiness Limitations, is approved by the FAA. Older aircraft maintenance manuals do not include an airworthiness limitations sections.

26. **Must all manufacturer's service letters, instructions, and bulletins be complied with on an annual or 100-hour inspection?** (AM.II.D.K7)
Not unless they have been incorporated into an airworthiness directive.

27. **What federal regulation stipulates the requirements for 100-hour and annual inspections?** (AM.II.D.K8)
14 CFR §91.409.

28. **How is an aircraft determined to be airworthy?** (AM.II.D.K8)
The airworthiness certificate of the aircraft states that the aircraft was found to be in conformance to its type certificate and in a condition for safe operation. Both of those criteria must continue to be true for the aircraft to be airworthy. In addition, the airworthiness certificate continues to be in effect as long as the aircraft is maintained in accordance with 14 CFR Parts 23, 43, and 91.

29. **What are the two general classifications of corrosion that cover most specific forms?** (AM.II.D.K9)
Direct chemical attack and electrochemical attack.

30. **What are the different types of corrosion and their identifying features?** (AM.II.D.K9)
    a. *Surface corrosion*—General roughening, etching, or pitting, frequently accompanied by a powdery deposit of corrosion products.
    b. *Filiform corrosion*—Worm-like traces of corrosion below the paint film.
    c. *Pitting corrosion*—First shows as a white or gray powdery deposit. When cleared away, tiny holes or pits can be seen in the surface.
    d. *Dissimilar metal corrosion*—Extensive pitting caused by the contact of two dissimilar metals in the presence of a conductor.
    e. *Concentration cell corrosion*—Metal-to-metal joint corrosion, or corrosion of a spot on the metal surface covered by a foreign material.
    f. *Intergranular corrosion*—An attack along the grain boundaries of an alloy.
    g. *Exfoliation corrosion*—An advanced form of intergranular corrosion that shows up as a lifting up of the surface grains of the metal by the expanding corrosion below the surface.

h. *Stress-corrosion/cracking*—Corrosion caused by a constant or cyclic stress acting in conjunction with a damaging chemical environment.
i. *Fretting corrosion*—A corrosive attack that occurs when two mating surfaces begin to have a slight relative motion between them.
j. *Fatigue corrosion*—When a cyclic stress and a corrosive environment combine to reduce the life of the part through pitting and cracking.
k. *Galvanic corrosion*—Corrosion that occurs when two dissimilar metals make electrical contact in the presence of an electrolyte.

Risk Management

1. **Why is it important to understand recommended inspection intervals?** (AM.II.D.R1)
   If an inspection interval is left to go too long, under maintenance occurs and can lead to failure of components because they were not serviced often enough. If an inspection interval happens too often, it may lead to premature wear of components due to repeated disassembly and assembly.

2. **What are some of the benefits and drawbacks of visual inspection?** (AM.II.D.R2)
   A good visual inspection is very useful and is often the only level of inspection needed. A visual inspection can be aided with a bright light, a magnifying lens, and a mirror. However, the lack of visible defects does not mean that further inspection is unnecessary. Defects that are below the surface or in nonvisible areas require additional inspection methods.

3. **What risks are associated with using radiographic inspection methods?** (AM.II.D.R3)
   Due to the radiation risks associated with x-ray, extensive training is required, and rigid safety measures must be followed.

4. **Why is it important to use the correct checklists and maintenance publications when performing aircraft maintenance?** (AM.II.D.R4)
   The use of incorrect checklists may lead to inspection items being missed or incorrect inspection processes. The use of incorrect maintenance publications may lead to incorrect procedures and processes, leading to unsafe maintenance practices.

5. **Why is accurate maintenance documentation important?** (AM.II.D.R5)
   Not only are maintenance records a legal requirement, they are critical for tracking of maintenance items and an understanding of the history and airworthiness of the aircraft.

Skills

1. **Determine from the aircraft records furnished by the examiner when the next 100-hour inspection is due, and when the next annual inspection is due.** (AM.II.D.S1)

2. **Determine from the aircraft records furnished by the examiner whether or not any repetitive Airworthiness Directives must be complied with on a 100-hour inspection.** (AM.II.D.S1)

3. **Using the aircraft model and serial number specified by the examiner, determine what Airworthiness Directives apply to the aircraft. Examine the aircraft maintenance records to determine if all of the applicable ADs have been complied with.** (AM.II.D.S1)

4. **Perform a conformity inspection on an engine, propeller, or airframe assigned by the examiner.** (AM.II.D.S1)

5. **Prepare an aircraft for a 100-hour inspection. Perform the inspection. Make the correct maintenance record entries to show that the inspection has been conducted.** (AM.II.D.S2)

6. **Describe to the examiner the record entry that must be made when an altimeter system has been inspected in accordance with 14 CFR §91.411 and 14 CFR Part 43, Appendices E and F.** (AM.II.D.S3)

7. **Explain to the examiner how to determine whether the given AD applies to the aircraft or component.** (AM.II.D.S4)

8. **Demonstrate to the examiner where to find a checklist for conducting a 100-hour inspection on a specified aircraft.** (AM.II.D.S5)

9. **Determine if any additional inspection requirements apply to a particular 100-hour inspection (special inspections, component requirements, filter elements, etc.).** (AM.II.D.S6)

10. **Conduct an inspection of a seat and seat belts and determine if the proper TSO requirements have been met.** (AM.II.D.S7)

E. Landing Gear Systems

References: AC 43.13-1; FAA-H-8083-31

Knowledge

1. **Where may the instructions for greasing a retractable landing gear on an aircraft be found?** (AM.II.E.K1)
   In the aircraft maintenance manual.

2. **How does a shimmy damper keep a nose wheel from shimmying?** (AM.II.E.K1)
   It acts as a small hydraulic shock absorber between the piston and the cylinder of the nose-wheel shock strut.

3. **What is the most likely cause of a landing gear warning system failing to warn when the landing gear is not down and locked?** (AM.II.E.K1)
   A faulty or misadjusted microswitch.

4. **What absorbs the taxi shocks in an oleo shock strut?** (AM.II.E.K2)
   The compressed air or nitrogen.

5. **How much air or nitrogen should be put into an oleo shock strut?** (AM.II.E.K2)
   With the strut serviced with fluid and the filler plug in place and the weight of the aircraft on the strut, put in enough air or nitrogen to extend the piston to the height specified in the aircraft maintenance manual.

6. **What is the purpose of using an exerciser jack when servicing a shock strut with fluid?** (AM.II.E.K2)
   The jack moves the piston up and down inside the cylinder to work all of the air out of the oil to be sure that the proper amount of oil is in the strut.

7. **What is a squat switch and where is it located?** (AM.II.E.K2)
   A landing gear safety switch that energizes a circuit to prevent the landing gear retraction handle from being moved to the RETRACT position when weight is on the landing gear. It is located in the torsion links of one of the main landing gears.

8. **What kind of device is normally used as a sensor to detect the condition of a retractable landing gear?** (AM.II.E.K2)
   A precision microswitch.

9. **What is the purpose of the centering cam in a nose-wheel shock strut?** (AM.II.E.K2)
   The centering cam forces the nose wheel straight back with the strut before it is retracted into the nose-wheel well.

10. **What absorbs the initial landing impact in an oleo shock strut?** (AM.II.E.K2)
    The transfer of oil from one compartment to another through a metered orifice.

11. **Where may a list of the approved lubricants for greasing an aircraft retractable landing gear be found?** (AM.II.E.K3)
    In the aircraft maintenance manual.

12. **Where can you find the specifications for the type of fluid used to service an aircraft landing gear shock strut?** (AM.II.E.K3)
    On the nameplate on the strut.

13. **How much oil should be put into an oleo shock strut?** (AM.II.E.K3)
    With the strut completely deflated, fill it to the level of the filler plug.

14. **On a bungee cord shock absorber, what visual indications depict a deteriorating bungee cord?** (AM.II.E.K4)
    Fraying of the outer cloth sheath and broken stands of elastic rubber.

15. **What should you look for when inspecting a spring steel landing gear?** (AM.II.E.K4)
    Inspect the spring steel for corrosion and damage. Verify that one of the gear legs is not bent by checking to verify that the aircraft wings are level. Inspect gear leg attachment bolts for condition and security. Inspect for cracks in the gear leg mounting structure.

16. **What is a common method of steering small aircraft nose wheels?** (AM.II.E.K5)
    A push-pull rod system attached to the rudder pedals.

17. **What method is used to steer large aircraft?** (AM.II.E.K5)
    Hydraulic steering systems.

18. **What would cause the warning horn to sound when the throttles are pulled back, reducing the engine power for landing?** (AM.II.E.K6)
    The warning horn will sound if any of the landing gears are not down and locked.

19. **What is indicated by a red light in the landing gear position-indication portion of the annunciator panel?** (AM.II.E.K6)
    The red light indicates that the landing gear is not in a safe condition for landing.

20. **What information is given to a pilot to indicate that all of the landing gears are down and locked?** (AM.II.E.K6)
    Three green lights are used on most aircraft to indicate that all three landing gears are down and locked.

21. **Where can you find a list of brake fluids approved for the aircraft?** (AM.II.E.K7)
    In the aircraft service manual.

22. **What is used to flush brake lines and cylinders if the system uses vegetable-base fluid?** (AM.II.E.K7)
    Alcohol.

23. **What is used to flush brake lines and cylinders if the system uses mineral-base fluid?** (AM.II.E.K7)
    Naphtha, varsol, or Stoddard solvent.

24. **How do you measure the amount of brake lining wear on single-disc brakes?** (AM.II.E.K7)
    Measure the clearance between the disc and the inboard side of the brake housing with the brakes applied.

25. **What is the purpose of the debooster in a hydraulic power brake system?** (AM.II.E.K7)
    The debooster decreases the pressure and increases the volume of fluid going to the brakes. This gives the pilot better control of the brakes.

26. **What should be done to hydraulic brakes when the pedal has a spongy feel?** (AM.II.E.K7)
    The spongy feel is caused by air in the brake. The brakes should be bled of this air.

27. **What is the purpose of the compensator port in the master cylinder of aircraft brakes?** (AM.II.E.K7)
    The compensator port in the master cylinder opens the brake reservoir to the wheel cylinders when the brakes are off. This prevents pressure from building up in the brake lines and causing the brakes to drag.

28. **What is the purpose of the shuttle valve in the brake system of an aircraft using hydraulic power brakes?** (AM.II.E.K7)

The shuttle valve is an automatic transfer valve. It allows the brakes to be operated by hydraulic system pressure under all normal conditions; but if this pressure is lost, it allows the brakes to be operated by the emergency backup system.

29. **How does an antiskid brake system keep the wheels of an aircraft from skidding on a wet runway?** (AM.II.E.K7)

The antiskid system monitors the rate of deceleration of the wheels. If any wheel slows down faster than it should (as it would at the beginning of a skid), the pressure on the brake in that wheel is released until the wheel stops decelerating, then the pressure is reapplied.

30. **What abnormal event warrants a special brake inspection?** (AM.II.E.K7)

If the aircraft experiences an aborted take-off or an emergency braking event occurs. These events cause excessive heating of brakes, wheels, and tires.

31. **What is the likely cause of "spongy brakes"?** (AM.II.E.K7)

There are air bubbles in the brake lines.

32. **What are the two methods of bleeding brake systems?** (AM.II.E.K7)

Gravity and pressure bleeding.

33. **What condition exists when a brake locks on a water covered runway?** (AM.II.E.K7)

Hydroplaning.

34. **What is the likely cause of a dragging brake?** (AM.II.E.K7)

A warped disc.

35. **What is the function of a wheel speed sensor in a anti-skid brake system?** (AM.II.E.K8)

The wheel speed sensor is a DC generator that is used to produce a voltage proportional to wheel speed. Rotational speed change in excess of programmed parameters will engage the anti-skid data.

36. **What is meant by the ply rating of a tire?** (AM.II.E.K9)

The number of plies of cotton fabric needed to produce the same strength as the actual plies in the tire.

37. **What is the function of the bead in a tire?** (AM.II.E.K9)

The bead is made of high-strength carbon steel wire bundles and used to provide the strength and stiffness where the tire mounts on the wheel.

38. **Is it proper to use a tube in a tubeless tire?** (AM.II.E.K9)

No, the inner liner of a tubeless tire is rough and it can chafe the tube in normal operation.

39. **What is the function of the grooves cut into the tread of a tire?** (AM.II.E.K9)

The grooves produce the optimum traction with the runway surface.

40. **What is the most widely used tread pattern for modern airplane tires?** (AM.II.E.K9)
A series of straight grooves around the periphery of the tire.

41. **How should new aircraft tires be stored?** (AM.II.E.K10)
In a dark, cool location away from electrical motors or battery chargers.

42. **Is it best to store new tires vertically or horizontally?** (AM.II.E.K10)
Vertically.

43. **What is the most important maintenance procedure for aircraft tires?** (AM.II.E.K10)
Keep them properly inflated.

44. **What should be used to remove oil from a tire?** (AM.II.E.K10)
Mild soap and warm water.

45. **Where do you find the proper inflation pressure for an aircraft tire?** (AM.II.E.K10)
In the aircraft service manual.

46. **Why should aircraft tire pressure be rechecked after it has been installed for about 24 hours with no load applied?** (AM.II.E.K10)
The tire will stretch and the pressure will drop. The proper pressure must be restored.

47. **What causes an aircraft tire to wear more on the shoulders than in the center of the tread?** (AM.II.E.K10)
Operating the tire in an underinflated condition.

48. **What causes an aircraft tire to wear more in the center of the tread than on the shoulders?** (AM.II.E.K10)
Operating the tire in an overinflated condition.

49. **What should be done to an aircraft tire if the sidewalls are weather-checked enough to expose the cord?** (AM.II.E.K10)
The tire should be scrapped.

50. **What should be done with a tire that was on a wheel which was overheated enough to melt the fusible plug in the wheel?** (AM.II.E.K10)
The tire should be scrapped.

51. **When removing a wheel from an aircraft to change the tire, when should the tire be deflated?** (AM.II.E.K11)
After the aircraft is on the jacks, but before the axle nut is loosened.

52. **What should be used to deflate high-pressure tires?** (AM.II.E.K11)
A deflator cap screwed onto the tire valve, to allow the air to escape through the hole in the side of the cap and not in the face of the mechanic.

53. **What is used in a split wheel to keep air from leaking between the two wheel halves?** (AM.II.E.K11)
An O-ring seal.

54. **Where are cracks most likely to form in an aircraft wheel?** (AM.II.E.K11)
In the bead seat area.

55. **What type of inspection should be used to inspect the bead seat area of a wheel for cracks?** (AM.II.E.K11)
Eddy current inspection.

56. **What are the three basic brake actuating systems?** (AM.II.E.K12)
    a. An independent system not part of the aircraft main hydraulic system.
    b. A booster system that uses the aircraft hydraulic system intermittently.
    c. A power brake system that only uses the aircraft main hydraulic system as a source of pressure.

57. **Where can you find instructions for hoisting an aircraft to replace wheels with floats?** (AM.II.E.K13)
In the aircraft maintenance manual.

Risk Management

1. **Where can you find instructions for jacking one wheel of an aircraft to change a tire?** (AM.II.E.R1)
In the aircraft maintenance manual.

2. **What safety procedure should be followed when inflating a tire on a large aircraft wheel for the first time after it as been replaced?** (AM.II.E.R1)
Put the wheel in a safety cage when it is being inflated because of the danger if the through bolts should fail.

3. **Why is it important that some aircraft with retractable landing gear be given a retraction test after new or retreaded tires are installed?** (AM.II.E.R1)
It is possible in some aircraft that a new or retreaded tire can be different enough in size from the previous tire that it could lock up in the wheel well when the landing gear is retracted.

4. **Where can you find safety precautions for jacking an aircraft for retractable landing gear inspection and maintenance?** (AM.II.E.R2)
In the aircraft maintenance manual.

5. **What precaution should be taken before working on a high-pressure fluid or gas systems?** (AM.II.E.R3)
Always depressurize a high-pressure system before disconnecting any lines or fittings and use appropriate PPE.

6. **Where are open containers of hydraulic fluid normally stored?** (AM.II.E.R4)
In a metal flammable chemicals storage cabinet.

7. **What are the risks of mixing hydraulic fluids?** (AM.II.E.R4)
Hydraulic fluids are not necessarily compatible with each other. Mixing of fluids may render fire-resistant fluid non-fire-resistant and may damage seals, causing system failure.

8. **What is the first step when taking apart a high-pressure strut or accumulator, and why is this important?** (AM.II.E.R5)

   Depressurize the system. Disassembly of a component that is pressurized can result in explosive depressurization, causing damage or injury.

9. **What risks are associated with operating retractable landing gear around personnel?** (AM.II.E.R6)

   When operating the landing gear retraction system from the flight deck, it is difficult or impossible to verify that personnel are clear of the moving landing gear.

Skills

1. **Demonstrate to the examiner the correct way to lubricate a retractable landing gear.** (AM.II.E.S1)
2. **Demonstrate to the examiner the correct way to jack one wheel of an aircraft, remove the wheel, disassemble the wheel, and remove the tire.** (AM.II.E.S1)
3. **Demonstrate the correct way to check and install the center O-ring, mount the tire on the wheel, torque the wheel nuts, and inflate the tire.** (AM.II.E.S1)
4. **Demonstrate the correct way to pack the wheel bearings with grease, install the wheel on the aircraft, adjust the axle nut, and safety it on the axle.** (AM.II.E.S1)
5. **Perform an operational check of a retractable landing gear and explain to the examiner the interconnection between the landing gear position switches, the engine throttle(s), and the landing gear warning horn.** (AM.II.E.S1)
6. **Identify the landing gear position system components.** (AM.II.E.S1)
7. **Identify the landing gear warning system components.** (AM.II.E.S1)
8. **Describe the sequence of operation for a landing gear warning system.** (AM.II.E.S1)
9. **Locate the troubleshooting procedures for a takeoff warning system.** (AM.II.E.S1)
10. **Troubleshoot the landing gear position and/or warning systems.** (AM.II.E.S1)
11. **Inspect the landing gear position indicating system.** (AM.II.E.S1)
12. **Repair the landing gear position indicating systems.** (AM.II.E.S1)
13. **Demonstrate the proper method to inspect, check, and service the supplied anti-skid system.** (AM.II.E.S2)
14. **Locate procedures for checking operation of an anti-skid warning system.** (AM.II.E.S3)

15. **Locate troubleshooting procedures for an anti-skid system.** (AM.II.E.S4)

16. **Demonstrate to the examiner the correct way to position the jacks under the wing of an airplane to raise it off the hangar floor. Explain the safety procedures that must be observed.** (AM.II.E.S5)

17. **Demonstrate to the examiner the correct way to jack an aircraft to perform a landing gear retraction test.** (AM.II.E.S5)

18. **Jack one wheel of an airplane so the wheel can be removed. Explain to the examiner the safety precautions that must be taken.** (AM.II.E.S5)

19. **Explain to the examiner the way a particular retractable landing gear is prevented from retracting when the weight of the aircraft is on it.** (AM.II.E.S6)

20. **Troubleshoot a retractable landing gear system.** (AM.II.E.S6)

21. **Make the proper maintenance record entry for the completion of a landing gear retraction test.** (AM.II.E.S6)

22. **Using an electrical schematic diagram of a landing gear warning system, explain to the examiner what fault could prevent the system from warning of an unsafe landing gear condition.** (AM.II.E.S6)

23. **Check the main wheels of an aircraft for the proper amount of camber, and toe-in or toe-out.** (AM.II.E.S7)

24. **Demonstrate your ability to check brake-lining wear and the condition of the disc on an aircraft specified by the examiner, and replace the lining.** (AM.II.E.S7)

25. **Demonstrate the correct way to inspect the wheel for cracks and corrosion.** (AM.II.E.S7)

26. **Demount a tire from an aircraft wheel, inspect the tire for wear, and explain to the examiner the conditions that could cause a tire to be unairworthy.** (AM.II.E.S7)

27. **Install a tire on a wheel and inflate it to the proper pressure. Explain to the examiner the safety procedures to follow.** (AM.II.E.S7)

28. **Repair a defective tire tube.** (AM.II.E.S7)

29. **Replace a tire or tube valve core and check for leaks.** (AM.II.E.S7)

30. **Inspect an electrical brake control for proper operation.** (AM.II.E.S7)

31. **Demonstrate to the examiner the removal and replacement of brake linings.** (AM.II.E.S8)

32. **Demonstrate to the examiner the correct way to service an oleo shock strut:**
    a. **Select the correct fluid.**
    b. **Explain the safety procedures to follow.**
    c. **Put in the correct amount of fluid.**
    d. **Put in the correct amount of air.**

    (AM.II.E.S9)

33. **Explain to the examiner the way to determine if an O-ring is the proper one for a particular application.** (AM.II.E.S9)

34. **Inspect a piece of shock absorber cord to determine if it has exceeded its shelf life for installation on a certificated aircraft.** (AM.II.E.S9)

35. **Demonstrate the correct way to bleed brakes on an aircraft specified by the examiner.** (AM.II.E.S10)

36. **Troubleshoot faults in the supplied hydraulic brake system.** (AM.II.E.S11)

37. **Demonstrate the correct way to replace the seals in the wheel cylinder of a single-disc brake.** (AM.II.E.S12)

38. **Demonstrate the correct way to remove, inspect, and re-install the provided wheel brake assembly.** (AM.II.E.S12)

39. **Inspect a tire provided by the examiner and determine if the tire is airworthy.** (AM.II.E.S13)

40. **Locate tire storage practices.** (AM.II.E.S14)

41. **Replace an air/oil shock strut air valve.** (AM.II.E.S15)

42. **Troubleshoot an air/oil shock strut.** (AM.II.E.S16)

43. **Inspect a nose-wheel shimmy damper, and service with fluid.** (AM.II.E.S17)

44. **Inspect and adjust the nose wheel steering supplied by the examiner.** (AM.II.E.S18)

45. **Troubleshoot a nose wheel steering system.** (AM.II.E.S18)

46. **Inspect the landing gear alignment.** (AM.II.E.S19)

47. **Locate and explain to the examiner the process for checking landing gear alignment on an assigned aircraft.** (AM.II.E.S19)

48. **Demonstrate to the examiner how to replace master brake cylinder packing seals on the supplied master cylinder.** (AM.II.E.S20)

49. **Demonstrate how to troubleshoot the aircraft steering system on the assigned aircraft.** (AM.II.E.S21)

50. **Identify the landing gear position and warning components on an aircraft.** (AM.II.E.S22)

51. **Locate procedures for checking pneumatic/bleed air overheat warning systems.** (AM.II.E.S23)

52. **Using a wiring diagram of a landing-gear warning system, explain to the examiner a malfunction that would prevent the system warning of an unlocked landing gear.** (AM.II.E.S23)

53. **Troubleshoot and repair the landing gear position and warning system on an aircraft assigned by the examiner.** (AM.II.E.S24)

54. **Demonstrate to the examiner the proper way to adjust landing gear position switches to the aircraft manufacturer's specifications.** (AM.II.E.S25)

55. **Remove, install, and/or adjust a landing gear down-lock switch.** (AM.II.E.S26)

56. **Check the rigging and adjustment of a landing gear up-lock.** (AM.II.E.S26)

57. **Demonstrate the correct way to check the fluid level in a brake master cylinder and determine the correct type of fluid.** (AM.II.E.S27)

58. **Inspect a brake system for serviceability.** (AM.II.E.S27)

59. **Inspect and troubleshoot a nose-wheel shimmy.** (AM.II.E.S28)

60. **Remove the tube from the supplied wheel assembly and inspect for damage and condition.** (AM.II.E.S29)

F. Hydraulic and Pneumatic Systems

References: AC 43.13-1; FAA-H-8083-31

Knowledge

1. **What kind of filter is a micronic filter?** (AM.II.F.K1)
   A filter with a special paper element.

2. **What is a double-action pump?** (AM.II.F.K1)
   A pump that delivers fluid with the movement of the pump handle in both directions.

3. **Why do most engine-driven hydraulic pumps have a shear section in their drive couplings?** (AM.II.F.K1)
   If the pump should seize, the shear section will break, disconnecting the pump from the engine and preventing further damage.

4. **What are the two basic types of hydraulic fluid used in modern aircraft?** (AM.II.F.K1)

   Mineral base fluid and phosphate ester base fluid.

5. **Does the main hydraulic pump take its fluid from the bottom of the reservoir, or from a standpipe?** (AM.II.F.K1)

   The main pump normally takes its fluid from a standpipe, while the emergency pump takes its fluid from the bottom of the reservoir. If a break in the system should allow the main pump to pump all of its fluid overboard, there will still be enough fluid in the reservoir to allow the emergency system to extend the landing gear and actuate the brakes.

6. **Why are some hydraulic reservoirs pressurized?** (AM.II.F.K1)

   Pressurization ensures that fluid will be supplied to the inlet of the pumps at high altitude where there is not enough atmospheric pressure to do this.

7. **What are two ways aircraft hydraulic reservoirs may be pressurized?** (AM.II.F.K1)

   By an aspirator in the fluid return line or by bleed air from one of the engine compressors.

8. **What is used to flush a hydraulic system that uses Skydrol hydraulic fluid?** (AM.II.F.K1)

   Trichlorethylene or the solvent recommended by the aircraft manufacturer.

9. **What is used to flush a hydraulic system that uses mineral base hydraulic fluid?** (AM.II.F.K1)

   Naphtha, varsol, or Stoddard solvent.

10. **Where can you find the type of hydraulic fluid required for a particular aircraft?** (AM.II.F.K1)

    In the maintenance manual for the aircraft. This information is also on a placard on the system reservoir.

11. **What type of pump is a gear pump considered?** (AM.II.F.K1)

    A gear pump is a constant displacement, high pressure pump.

12. **What type of pump is a piston pump with a variable cam plate considered?** (AM.II.F.K1)

    Fuel pumps may be divided into two distinct system categories: (1) constant displacement and (2) variable displacement. A piston pump with a variable cam changes the effective (not actual) length of each piston stroke. This makes this type of pump a variable displacement with constant pressure.

13. **What is the purpose of an accumulator in an aircraft hydraulic system?** (AM.II.F.K2)

    The accumulator holds pressure on the hydraulic fluid in the system. The pressure is held by compressed air or nitrogen acting on the fluid through a bladder, a diaphragm, or a piston.

14. **What is a single-action hydraulic actuating cylinder?** (AM.II.F.K2)

    A linear actuating cylinder that uses hydraulic fluid under pressure to move the piston in only one direction. The piston is returned by a spring.

15. **What is the difference between a single-acting and double-acting hand pump?** (AM.II.F.K2)

    Single-action pumps move fluid on only one direction of handle operation. Double-acting hand pumps move fluid on both directions of handle movement.

16. **If a system requires a lower pressure than system pressure, what type of valve is used?** (AM.II.F.K2)

    A pressure reducer valve.

17. **What is used to remove phosphate-ester base hydraulic fluid from aircraft tires?** (AM.II.F.K3)

    Soap and water.

18. **How can you be sure that a replacement seal in a hydraulic component is compatible with the fluid used in the system?** (AM.II.F.K3)

    Use only the seal identified by part number that is specified in the component maintenance manual.

19. **What must be done to the lines that are disconnected when replacing an aircraft hydraulic system component?** (AM.II.F.K3)

    They must be capped with the correct fluid line cap or plug. Masking tape or other types of adhesive tape should never be used.

20. **What must be done before a pressurized reservoir is serviced?** (AM.II.F.K3)

    All of the pressure must be bled off.

21. **What system fault could cause an accumulator pre-charge pressure to increase?** (AM.II.F.K4)

    A leaking seal in the accumulator that allows fluid to pass into the air side of the accumulator.

22. **What is the first step when checking hydraulic reservoir fluid level?** (AM.II.F.K4)

    Verify that the hydraulic systems are set to the configuration specified in the maintenance manual. For example: landing gear down, gear doors retracted, parking brake off, thrust reversers retracted, and pressure accumulators set to the correct pressures.

23. **What is the most likely cause of hydraulic fluid flowing through a filter bypass valve?** (AM.II.F.K4)

    The filter has become clogged.

24. **What are two basic differences between a hydraulic and a pneumatic system?** (AM.II.F.K5)

    The fluid in a pneumatic system is compressible and that used in a hydraulic system is incompressible. A pneumatic system has no return system.

25. **What are two functions of a low-pressure pneumatic system in an aircraft?** (AM.II.F.K5)

    To operate the gyros in flight instruments, and to inflate pneumatic deicer boots.

26. **What type of pump is used in a low-pressure pneumatic system?** (AM.II.F.K5)
A vane-type pump.

27. **Why do high-altitude aircraft that have pneumatic gyro instruments use pressure-actuated instruments rather than vacuum-actuated instruments?** (AM.II.F.K5)
At high altitude there is not enough air mass to drive the gyros at their required speed.

28. **What effect would a dirty inline air filter have on the operation of pneumatic gyro instruments?** (AM.II.F.K5)
The instruments would not get enough air to spin up to their required speed.

29. **Of what material are the vanes in a dry type air pump made?** (AM.II.F.K5)
Carbon.

30. **What kind of device is used to control the speed of movement of the piston in a pneumatic actuator?** (AM.II.F.K5)
A variable orifice.

31. **What is the purpose of the moisture separator in a high-pressure pneumatic system?** (AM.II.F.K5)
The separator collects moisture from the air and holds it on a baffle until the system is shut down. When the inlet pressure to the moisture separator drops below 450 psi, a drain valve opens and all of the accumulated moisture is discharged overboard. A moisture separator removes about 98% of the moisture in the air.

32. **What are two types of filters that are used in an aircraft pneumatic system?** (AM.II.F.K5)
Micronic (paper) type, and screen type.

33. **Where is the inlet air filter normally located for a pneumatic gyro instrument system powered by engine-driven air pumps?** (AM.II.F.K6)
Inside the cabin of the aircraft.

34. **Where is the source of air when performing an operational check of a pneumatic deice boot system?** (AM.II.F.K6)
It is provided either by operating the aircraft engines, or from an external source of air through pressure-regulated test equipment.

35. **How is the air charge of an accumulator checked for proper servicing?** (AM.II.F.K7)
All hydraulic pressure must be bled from the system and the pressure is checked on the gauge.

36. **Is a chevron seal a one-way or a two-way seal?** (AM.II.F.K8)
A one-way seal.

37. **To which side of a chevron seal should the pressure be applied?** (AM.II.F.K8)
To the open side.

38. **Is an O-ring seal a one-way or a two-way seal?** (AM.II.F.K8)
A two-way seal.

39. **How can you be sure of getting an O-ring made of the correct material for a specific hydraulic component?** (AM.II.F.K8)
The O-ring must have the correct part number and it must have been obtained from a reputable source.

40. **On which side of an O-ring should the backup ring be placed?** (AM.II.F.K8)
On the side away from the source of pressure.

41. **What should be done to the sharp edges of threads and actuator pistons when installing O-rings?** (AM.II.F.K8)
The sharp edges should be covered with paper.

42. **Where are line-disconnect fittings normally located in an aircraft hydraulic system?** (AM.II.F.K9)
Normally in the lines that connect the engine-driven pump to the aircraft hydraulic system.

43. **What is the purpose of an orifice check valve in an aircraft hydraulic system?** (AM.II.F.K10)
An orifice check valve allows full flow of fluid in one direction through the valve, but restricts the flow in the opposite direction.

44. **What is the difference in function between a pressure regulator and a pressure relief valve?** (AM.II.F.K10)
Pressure regulators control overall system pressure. Pressure relief valves act a back-up to the pressure regulator to prevent overpressure damage to the system and components.

45. **What does an unloading valve do in a hydraulic system?** (AM.II.F.K10)
The unloading valve, or pressure regulator, controls system pressure by shifting the pump outlet fluid from the pressurized system back into the reservoir when the system pressure is high enough. The fluid circulates with very little load on the pump until the system pressure drops to the regulator kick-in value. The pump then forces fluid into the system until the pressure builds back up to the regulator kick-out value.

46. **What is a sintered metal hydraulic filter?** (AM.II.F.K11)
A surface filter whose element is made of powdered metal fused under heat and pressure.

47. **What is the purpose of the differential pressure indicator on a hydraulic filter?** (AM.II.F.K11)
It indicates when the filter is clogged and fluid is bypassing the filter element.

48. **What is a Cuno filter?** (AM.II.F.K11)
An edge filter made of a stack of thin metal discs separated by thin scraper blades. Contaminants collect on the edge of the discs and are periodically scraped out by rotating the discs. The contaminants collect in the bottom of the filter case for future removal.

Risk Management

1. **What is the danger of not relieving hydraulic system pressure before servicing or disassembly?** (AM.II.F.R1)
   High-pressure fluid can be released, causing damage to aircraft systems or personnel.

2. **What precautions should be taken when working around high-pressure gases and fluids?** (AM.II.F.R2)
   Wear the proper PPE and always depressurize the system in accordance with the aircraft maintenance manual prior to any disassembly.

3. **Where should hydraulic fluids be stored?** (AM.II.F.R3)
   Many hydraulic fluids are flammable and should be stored in metal, flammable-resistant cabinets.

4. **What PPE should be considered when working with hydraulic fluids?** (AM.II.F.R3)
   Always consult the Safety Data Sheet (SDS) for the fluids. Eye protection and skin protection (gloves) are minimum PPE.

5. **What are the dangers of mixing hydraulic fluids?** (AM.II.F.R4)
   The desired properties of the fluids will be compromised. Damage may occur to internal O-rings and seals, causing systems to fail.

6. **What is the risk of using the wrong seal or hydraulic fluid in a system?** (AM.II.F.R5)
   Incompatible seals and hydraulic fluids will cause seals to deteriorate and fail, leading to a failure of the system.

Skills

1. **Identify different types of hydraulic fluids provided by the examiner.** (AM.II.F.S1)

2. **Identify the type and use of different packing seals.** (AM.II.F.S2)

3. **Demonstrate the correct way to select and install an O-ring seal in a hydraulic actuator.** (AM.II.F.S3)

4. **Remove and install a hydraulic selector valve.** (AM.II.F.S4)

5. **Check a pressure regulator for performance and adjust as necessary.** (AM.II.F.S5)

6. **Demonstrate to the examiner the correct way to install a filter in a low-pressure pneumatic system.** (AM.II.F.S5)

7. **Inspect a hydraulic system filter and service it according to the instructions of the aircraft manufacturer.** (AM.II.F.S6)

8. **Check the contaminants in a hydraulic filter and explain to the examiner the probable source of the contaminants and what action, if any, should be taken.** (AM.II.F.S6)

9. **Remove, clean, inspect, and install a hydraulic system filter.** (AM.II.F.S6)

10. **Service a hydraulic accumulator.** (AM.II.F.S7)

11. **Service the emergency air bottle in an aircraft brake system with the correct amount of air or nitrogen.** (AM.II.F.S7)

12. **Demonstrate to the examiner the correct way to service a pressurized hydraulic reservoir.** (AM.II.F.S8)

13. **Remove, install, and check an engine-driven hydraulic pump.** (AM.II.F.S9)

14. **Demonstrate to the examiner where to find the procedures for checking pneumatic/bleed air overheat warnings on the assigned system.** (AM.II.F.S10)

15. **Purge air from a hydraulic system.** (AM.II.F.S11)

16. **Adjust a pneumatic power system relief valve.** (AM.II.F.S12)

17. **Remove and/or install a system pressure relief valve.** (AM.II.F.S12)

18. **Inspect a hydraulic or pneumatic system for leaks.** (AM.II.F.S13)

19. **Troubleshoot a pneumatic power system leak.** (AM.II.F.S14)

20. **Troubleshoot hydraulic power system leaks.** (AM.II.F.S14)

21. **Troubleshoot the operation of a hydraulic power system.** (AM.II.F.S14)

22. **Determine the correct type of hydraulic fluid required by a particular aircraft and check the system for the correct amount of fluid.** (AM.II.F.S15)

23. **Identify proper installation procedures for a seal, backup ring, and/or gasket.** (AM.II.F.S16)

G. Environmental Systems

References: AC 43.13-1; FAA-H-8083-31

Knowledge

1. **Why are the cabins of most turbine-powered aircraft pressurized?** (AM.II.G.K1)
   These aircraft fly at such high altitudes that supplemental oxygen would be needed for the occupants if the cabins were not pressurized.

2. **Where does the pressurizing air come from on most turbine-powered aircraft?** (AM.II.G.K1)
From air bled from one of the engine compressors.

3. **Where does the pressurizing air come from on most smaller reciprocating-engine-powered aircraft?** (AM.II.G.K1)
From the engine turbocharger.

4. **What determines the amount of pressurization that an aircraft can use?** (AM.II.G.K1)
The structural strength of the aircraft cabin.

5. **What is meant by the isobaric mode of cabin pressurization?** (AM.II.G.K1)
The isobaric mode of cabin pressurization is the mode that keeps the cabin altitude constant as the aircraft changes its flight altitude.

6. **What is meant by the constant differential mode of cabin pressurization?** (AM.II.G.K1)
After the pressure in the aircraft cabin reaches the maximum value allowed by structural considerations, the constant differential mode of operation holds the pressure inside the cabin a constant amount above the outside air pressure.

7. **How is high-temperature bleed air cooled down to safe temperatures for aircraft heating?** (AM.II.G.K2)
The air is passed through a heat exchanger.

8. **How are aircraft instruments and avionics cooled?** (AM.II.G.K3)
By routing air-conditioned air over the instruments and avionics, or by electrically operated cooling fans.

9. **What is the danger of a leak in an exhaust system cabin heater?** (AM.II.G.K4)
Exhaust gases contain carbon monoxide (CO), a colorless, odorless, gas that can cause incapacitation or death to the aircraft occupants.

10. **How can an exhaust system be checked for leaks?** (AM.II.G.K4)
Pressurize it with the discharge from a vacuum cleaner and wipe the outside of the system with a soap solution. Any leak will cause bubbles to form.

11. **How can you tell that there is carbon monoxide in the aircraft cabin?** (AM.II.G.K4)
Indicator crystals in a CO detector will change color, from a normally bright color to a dark color. At lethal levels of CO they will turn black. Electronic CO detectors provide both visual and aural alerts to the crew.

12. **Where does the warm air come from that is used to heat the cabin of most small single-engine reciprocating-engine-powered aircraft?** (AM.II.G.K4)
From a shroud around the engine muffler.

13. **Where does the fuel used in an aircraft combustion heater come from?** (AM.II.G.K5)
From the aircraft fuel tanks.

14. **What happens to a combustion heater if the flow of ventilating air is restricted?** (AM.II.G.K5)
If the ventilating air is restricted and the temperature reaches a preset value, the limit switch will cause the fuel to be shut off to the heater.

15. **What regulates the temperature in an aircraft cabin that is heated with a combustion heater?** (AM.II.G.K5)
A thermostat senses the cabin temperature and cycles the fuel valve on or off to maintain the temperature at the desired value.

16. **What maintenance is required for a combustion heater?** (AM.II.G.K5)
Clean the fuel filters and check for fuel leaks. Pressure test the combustion chamber to check for leaks.

17. **How is the heat removed from an aircraft cabin with a vapor-cycle air conditioning system?** (AM.II.G.K6)
The cabin heat is absorbed by the refrigerant in the evaporator, then carried outside the aircraft where it is given up to the outside air in the condenser.

18. **What produces the cool air in a vapor-cycle air conditioning system?** (AM.II.G.K6)
Warm cabin air is blown across the evaporator, where its heat is transferred into the refrigerant. The air that leaves the evaporator is cool.

19. **What is used as the refrigerant in a vapor-cycle air conditioning system?** (AM.II.G.K6)
A Freon-type liquid refrigerant known as Refrigerant 12, or the more environmentally friendly R-134a.

20. **What is the state of the refrigerant as it leaves the compressor?** (AM.II.G.K6)
It is a high-pressure gas.

21. **What is the state of the refrigerant as it leaves the condenser?** (AM.II.G.K6)
It is a high-pressure liquid.

22. **What is the state of the refrigerant as it leaves the thermostatic expansion valve?** (AM.II.G.K6)
It is a low-pressure liquid.

23. **What is the state of the refrigerant as it leaves the evaporator?** (AM.II.G.K6)
It is a low-pressure vapor.

24. **How is the compressor lubricated?** (AM.II.G.K6)
Refrigeration oil is mixed with the refrigerant and it circulates through the compressor and the entire system.

25. **What is the function of the thermostatic expansion valve?** (AM.II.G.K6)
It meters just enough liquid refrigerant into the evaporator that all of it will be evaporated by the time it leaves the evaporator coils.

26. **How is refrigerant put into the system?** (AM.II.G.K6)
It is put into the system through service valves on the low side of the system using a manifold gauge set.

27. **How should the components in the low side of a properly operating vapor-cycle cooling system feel?** (AM.II.G.K6)
They should feel cool.

28. **How is a vapor-cycle cooling system checked for refrigerant leaks?** (AM.II.G.K6)
Hold the probe of an electronic leak detector below any suspect fitting or component. If there is a leak, the tone of the sound produced by the detector will change.

29. **What safety equipment should be worn when charging a vapor-cycle cooling system?** (AM.II.G.K6)
A safety face shield that protects the entire face.

30. **What are two types of air conditioning systems that may be installed on an aircraft?** (AM.II.G.K6)
Air-cycle systems and vapor-cycle systems.

31. **Where does the warm air come from that is used to heat the cabin of a large jet transport aircraft?** (AM.II.G.K7)
Warm engine compressor bleed air is used.

32. **Where is the first place the hot compressor bleed air gives up some of its heat in an air-cycle cooling system?** (AM.II.G.K7)
In the primary heat exchanger.

33. **What is the function of the air-cycle machine in an air-cycle cooling system?** (AM.II.G.K7)
The centrifugal compressor increases the pressure and temperature of the bleed air. This high temperature air gives up some of its heat in the secondary heat exchanger, and a great deal more as it drives the expansion turbine. It leaves the expansion turbine as cold air.

34. **Why must air-cycle air conditioning systems incorporate a water separator?** (AM.II.G.K7)
The rapid cooling of the air in the expansion turbine causes moisture to condense in the form of fog. This moisture is trapped in the moisture separator before the air is released into the cabin.

35. **How is the temperature of the air produced by an air-cycle cooling system controlled?** (AM.II.G.K7)
By a temperature control valve which mixes hot engine compressor bleed air with cold air from the expansion turbine.

36. **How is cabin pressure controlled in a pressurized aircraft?** (AM.II.G.K8)
More pressure than is needed is pumped into the aircraft cabin, and the pressure controller modulates the outflow valve to maintain the correct pressure in the cabin.

37. **What is the function of the cabin outflow valve on a pressurized aircraft?** (AM.II.G.K8)
The cabin outflow valve, which is controlled by the pressure controller, maintains the correct amount of pressure inside the cabin.

38. **What is the function of the cabin pressure safety valve on a pressurized aircraft?** (AM.II.G.K8)
The cabin pressure safety valve prevents cabin pressure from exceeding the maximum allowable differential pressure.

39. **Why must pressurized aircraft have a negative pressure relief valve?** (AM.II.G.K8)
The structure of an aircraft cabin is not designed to tolerate the inside pressure being lower than the outside pressure.

40. **What keeps the cabin of a pressurized aircraft from being pressurized when the aircraft is on the ground?** (AM.II.G.K8)
A squat switch on the landing gear holds the safety valve open when the aircraft is on the ground.

41. **What are three ways supplemental oxygen can be carried in an aircraft?** (AM.II.G.K9)
As a high-pressure gas, in its liquid form, and as a solid in the form of a chemical candle.

42. **What is a continuous-flow oxygen system?** (AM.II.G.K9)
An oxygen system that continuously flows a metered amount of oxygen into the mask.

43. **What is a pressure-demand oxygen system?** (AM.II.G.K9)
An oxygen system that flows oxygen to the mask only when the wearer of the mask inhales. Above a specified altitude, the regulator meters oxygen under pressure into the mask when the wearer inhales.

44. **What are the two main gases that make up our atmosphere?** (AM.II.G.K10)
Nitrogen and oxygen.

45. **What is used to check an oxygen system for leaks?** (AM.II.G.K10)
A special leak-detector liquid that is a form of non-oily soap.

46. **Why must fittings in an oxygen system not be tightened to stop a leak when there is pressure on the system?** (AM.II.G.K10)
When the system is pressurized, the tubing is expanded slightly; if the fitting is tightened when it is expanded, it will likely leak when the pressure is reduced.

47. **What kind of gaseous oxygen must be used to service an aircraft oxygen system?** (AM.II.G.K10)
Only aviators' breathing oxygen. Hospital oxygen and welding oxygen contain too much moisture to be used.

48. **What identification must be stamped on an oxygen bottle carried in an aircraft?** (AM.II.G.K10)
The identification DOT 3AA or DOT 3HT, the date of manufacture, and the date of all of the hydrostatic tests.

49. **To what pressure and how often should DOT 3AA oxygen cylinders be hydrostatically tested?** (AM.II.G.K10)
They should be tested to 5/3 of their working pressure every five years.

50. **To what pressure and how often should DOT 3HT oxygen cylinders be hydrostatically tested, and when should they be retired from service?** (AM.II.G.K10)
They should be tested to 3,083 psi every three years and retired from service after 15 years or 4,380 pressurizations, whichever occurs first.

51. **What kind of lubricant can be used for installing fittings in an oxygen system component?** (AM.II.G.K10)
Teflon tape or a special water-base lubricant.

52. **What is used to check for leaks after replacing a fitting in an oxygen system?** (AM.II.G.K10)
A special leak-detector liquid that is a form of non-oily soap.

53. **What cleaning solutions can be used to clean parts used in an oxygen system?** (AM.II.G.K10)
Anhydrous ethyl alcohol, isopropyl alcohol, or freon.

54. **What may be used to dry components in an oxygen system after they have been cleaned?** (AM.II.G.K10)
Water-pumped dry nitrogen.

55. **What may be used to purge the lines in an oxygen system after it has been opened for servicing?** (AM.II.G.K10)
Water-pumped dry nitrogen.

56. **What must be done before any maintenance can be done on an oxygen system?** (AM.II.G.K10)
The oxygen supply must be turned off at the bottle valve.

Risk Management

1. **What cautions should be used when performing maintenance on oxygen systems?** (AM.II.G.R1)
Follow the instructions in the aircraft maintenance manual, never allow oil on the threads of attaching components, use caution when working around high pressure gas, and use appropriate PPE.

2. **Why is it important to use refrigerant recovery equipment when servicing vapor-cycle air-conditioning systems?** (AM.II.G.R2)
Refrigerant has been shown to cause damage to the environment and should not be vented into the air.

3. **What cautions are to be used when handling, storing, or shipping chemical oxygen generators?** (AM.II.G.R3)

   Once lit, the generators cannot be extinguished and they produce heat. When not installed in an aircraft, they must be protected from accidental actuation with safety pins and protective covers. They are considered hazardous materials for shipping.

4. **What precautions should be taken for the storage, handling, and use of compressed gas cylinder and high-pressure systems?** (AM.II.G.R4)

   High-pressure cylinders should be stored standing vertically and retained with a strap or chain to prevent them from falling over. A protective cap should be used to cover the valve when not in use. Use appropriate high-pressure cylinder carts and dollies when moving cylinders and never lift a cylinder by the valve or protective cap. Use the correct PPE and follow manufacturer's instructions when servicing aircraft.

5. **What is the best way to mitigate risks associated with servicing vapor-cycle refrigerant systems?** (AM.II.G.R5)

   Follow the aircraft maintenance manual for servicing instructions, proper equipment and tooling, and warnings and cautions provided in the manual.

6. **What are some of the dangers associated with maintenance of a combustion heater?** (AM.II.G.R6)

   Failure to follow the maintenance instructions could lead to carbon monoxide entering the cabin, or the unit catching on fire.

Skills

1. **Inspect an oxygen system and demonstrate to the examiner the correct way to check an oxygen system for leaks.** (AM.II.G.S1)

2. **Demonstrate to the examiner the correct way to purge an oxygen system to remove all traces of air from the lines.** (AM.II.G.S2)

3. **Demonstrate to the examiner the correct way to service an installed oxygen system.** (AM.II.G.S3)

4. **Check the oxygen bottles in an aircraft for the required identification marks and for the status of their hydrostatic tests.** (AM.II.G.S3)

5. **Demonstrate to the examiner the correct way to remove and install an oxygen valve or fitting.** (AM.II.G.S3)

6. **Demonstrate to the examiner the correct way to service an oxygen system with the proper type and amount of oxygen.** (AM.II.G.S3)

7. **Demonstrate to the examiner the correct way to check an oxygen system for leaks.** (AM.II.G.S3)

8. **Demonstrate to the examiner how to inspect and clean a pilot emergency oxygen mask and supply hoses.** (AM.II.G.S4)

9. **Demonstrate to the examiner how to inspect an oxygen system pressure regulator.** (AM.II.G.S5)

10. **Demonstrate to the examiner how to inspect an oxygen system cylinder for serviceability.** (AM.II.G.S6)

11. **Demonstrate to the examiner how to inspect a chemical oxygen generator for serviceability and safe handling.** (AM.II.G.S7)

12. **Demonstrate to the examiner how to inspect a combustion heater fuel system for leaks.** (AM.II.G.S8)

13. **Demonstrate to the examiner where to find the procedures for troubleshooting a combustion heater.** (AM.II.G.S8)

14. **Given a schematic diagram of a vapor-cycle cooling system, explain the function of each of the components to the examiner.** (AM.II.G.S9)

15. **Locate the operating instructions for a Freon system.** (AM.II.G.S9)

16. **Locate the procedures for servicing a refrigerant (vapor-cycle) system.** (AM.II.G.S9)

17. **Demonstrate to the examiner the correct way to check a vapor-cycle cooling system for refrigerant leaks.** (AM.II.G.S9)

18. **Inspect a combustion heater in an aircraft; explain to the examiner the way the system operates and the precautions that should be taken in its operation.** (AM.II.G.S10)

19. **Locate the troubleshooting procedures for an air-cycle system.** (AM.II.G.S11)

20. **Locate the sources of contamination in a Freon system.** (AM.II.G.S12)

21. **Given a schematic diagram of an air-cycle cooling system, explain to the examiner the function of each of the components.** (AM.II.G.S12)

22. **Explain to the examiner how to troubleshoot the supplied air-cycle air conditioning system.** (AM.II.G.S12)

23. **Inspect for leaks in the portion of the engine exhaust system that supplies heat for the aircraft cabin heater.** (AM.II.G.S13)

24. **Given a schematic diagram of an aircraft pressurization system, explain to the examiner the function of each of the components.** (AM.II.G.S14)

25. **On a pressurized aircraft specified by the examiner, locate and identify the source of pressurizing air, the cabin outflow valve, the cabin pressure safety valve, and the cabin pressure controller.** (AM.II.G.S14)

26. **Locate the instructions for the inspection of a pressurization system.** (AM.II.G.S14)

27. **Demonstrate to the examiner where to find the troubleshooting procedures for a pressurization system.** (AM.II.G.S15)

H. Aircraft Instrument Systems

References: 14 CFR Parts 43, 91; AC 25-11, AC 43.13-1, AC 43-215; FAA-H-8083-31

Knowledge

1. **Where are the system descriptions found for annunciator indications and the meaning of the associated warnings, cautions, and advisory lights?** (AM.II.H.K1)
   In the pilot's operating handbook (POH)/airplane flight manual (AFM) and the aircraft instructions for continued airworthiness (ICAs).

2. **What fluid is used in an aircraft magnetic compass?** (AM.II.H.K2)
   A special water-clear fluid similar to kerosine.

3. **What is the maximum amount of deviation error allowed when a magnetic compass is installed in an aircraft?** (AM.II.H.K2)
   10 degrees.

4. **Where should the compass correction card be placed?** (AM.II.H.K2)
   In plain sight of the pilot, near the compass.

5. **What is the difference between variation and deviation in a compass system?** (AM.II.H.K2)
   Compass variation is caused by the difference between the earth's magnetic and geographic poles. Aeronautical charts display magnetic variation. Compass deviation is caused by all other magnetic fields aside from the earth's magnetic and geographic poles. Deviation may be caused by magnetic fields in the aircraft.

6. **What error is corrected when an aircraft compass is swung?** (AM.II.H.K3)
   Deviation error.

7. **What is done to a compass to correct for deviation error?** (AM.II.H.K3)
   The compensating magnets in the compass are adjusted to minimize the effect of outside magnetic fields.

8. **Where on an airport is a compass rose located?** (AM.II.H.K3)
   At a location where there is little traffic and the area is free from magnetic interference caused by electrical power lines or buried pipes.

9. **What are the three fundamental pressure-sensing mechanisms used in aircraft instrument systems?** (AM.II.H.K4)
   The Bourdon tube, the diaphragm or bellows, and the solid-state sensing device.

10. **What basic mechanism is used to measure oil, fuel, and oxygen pressures?** (AM.II.H.K4)
    The Bourdon tube.

11. **What mechanical means is used to measure pressure in an airspeed indicator?** (AM.II.H.K4)
    A diaphragm.

12. **A failure of an airspeed indicator needle to move is likely caused by what problem?** (AM.II.H.K4)
    A blockage in the pitot head will prevent the creation of a differential pressure.

13. **What mechanical means can be used to measure oil temperature?** (AM.II.H.K5)
    A fluid filled, sealed system comprising of a bulb and capillary tube connected to a Bourdon tube. As temperature increases, the volume of the fluid increases and expands the Bourdon tube, which is connected to an indicator dial.

14. **How does a thermocouple measure temperature?** (AM.II.H.K5)
    The unit is comprised of two unlike metals. As temperature increases, a small voltage is produced and sent to a temperature indicator. The higher the temperature, the higher the voltage.

15. **Why is the length of a thermocouple lead important?** (AM.II.H.K5)
    Thermocouple accuracy is dependent on the resistance of the circuit. Lead lengths are designed for specific installations and should never be altered.

16. **What type of electrical circuit is used to measure small changes of resistance in an electrical resistance temperature probe?** (AM.II.H.K5)
    A Wheatstone-bridge circuit.

17. **What type of system is used to indicate the position of the wing flaps?** (AM.II.H.K6)
    Usually a resistance-type remote-indicating system such as the DC Selsyn system.

18. **What information is shown by the wing flap position indicator?** (AM.II.H.K6)
    The number of degrees the flaps are lowered.

19. **What are the gyroscopic instruments that are connected to the low-pressure pneumatic system of an aircraft?** (AM.II.H.K7)
    Heading indicator, attitude indicator, and turn-and-slip indicator.

20. **What device is used with a wet-pump vacuum system to prevent oil from getting into the deicer boots?** (AM.II.H.K7)
    An oil separator.

21. **What are two types of filters used with a pressure system for gyros?** (AM.II.H.K7)
    A pump inlet filter and an inline filter.

22. **What type of filter is used with a vacuum system for gyros?** (AM.II.H.K7)
A central air filter.

23. **What may be causing an attitude indicator to not properly erect or display?** (AM.II.H.K7)
Gyroscopes are dependent on achieving and maintaining proper rotational speed. Failure to do so may be caused by low vacuum created by the vacuum system or motor failure in an electrically-driven gyroscope.

24. **What type of direction-finding compass has no moving parts and easily integrates with digital systems?** (AM.II.H.K8)
Solid state magnetometers.

25. **How does a flux gate compass operate?** (AM.II.H.K8)
The flux gate compass consists of a very magnetically permeable circular segmented core frame called a spider. The earth's magnetic field flows through this iron core and varies its distribution through segments of the core as the flux valve is rotated via the movement of the aircraft. Pickup coil windings are located on each of the core's spider legs that are positioned 120° apart. The signal from the flux gate is passed to the flight deck via an autosyn transmitter.

26. **What two types of pumps are used in the low-pressure pneumatic system?** (AM.II.H.K9)
A dry-type and a wet-type air pump.

27. **What are the vanes of a wet-type air pump made of?** (AM.II.H.K9)
Steel.

28. **What are the vanes of a dry-type air pump made of?** (AM.II.H.K9)
Carbon.

29. **How are the vanes of a wet-type air pump lubricated?** (AM.II.H.K9)
With engine oil directed into the pump through a small hole in the base of the pump.

30. **Why is it important to use the correct gasket when replacing a wet-type air pump?** (AM.II.H.K9)
The gasket must have a hole through which the pump lubricating oil can flow.

31. **Why do dry-type air pumps not need to be lubricated?** (AM.II.H.K9)
The special carbon material of which the vanes are made wears away in microscopic amounts to provide the needed lubrication.

32. **What must be done to the gyro pressure system if a dry-type air pump fails?** (AM.II.H.K9)
All of the filters must be replaced to prevent any debris from the pump vanes getting into the gyro instruments.

33. **Who is authorized to perform the altimeter tests to determine the accuracy of the altimeter?** (AM.II.H.K10)
The manufacturer of the aircraft on which the tests are conducted, or a certificated repair station properly equipped and authorized to perform the test.

34. **Where are the requirements for the altimeter system tests found?** (AM.II.H.K10)
14 CFR Part 43, Appendix E.

35. **How often should an altimeter be checked if it is installed in an aircraft used in IFR flight?** (AM.II.H.K10)
Every 24 calendar months.

36. **How much difference is allowed between the altitude indication on the automatic pressure altitude reporting equipment and that on the altimeter?** (AM.II.H.K10)
125 feet.

37. **To what altitude must altimeters be tested?** (AM.II.H.K10)
To the highest altitude the aircraft will be flown on IFR flight.

38. **What record must be made of a test of an altimeter?** (AM.II.H.K10)
The aircraft permanent maintenance record must show the date, the maximum altitude to which the altimeter was tested, and the name of the person approving the aircraft for return to service after the test.

39. **To what avionic equipment is the output from an encoding altimeter connected?** (AM.II.H.K10)
The ATC transponder.

40. **What is the allowable difference between the surveyed elevation of the airport and the indication on the altimeter when it is set to the local altimeter setting?** (AM.II.H.K10)
75 feet.

41. **What instruments in an aircraft are connected to the static system?** (AM.II.H.K10)
The airspeed indicator, vertical speed indicator, and altimeter.

42. **What are the differences between absolute, gauge, and differential pressure instrument systems?** (AM.II.H.K10)
An absolute pressure instrument, such as an altimeter, measures actual pressure. Gauge pressure instruments measure the difference between exiting barometric pressure and pressure inside a sealed container such as a Bourdon tube or a bellows. Differential pressure instruments display the difference between two pressures, such as pitot pressure and static pressure used to display airspeed.

43. **What are the basic methods of measuring fuel quantity in an aircraft?** (AM.II.H.K11)
Direct reading systems (sight gauges), floats that actuate a variable resistor, and capacitance-type systems.

44. What do the green, yellow, and red arc markings on an instrument mean? (AM.II.H.K12)

Green indicates normal operating range, yellow indicates caution operating range, and red indicates prohibited operating range.

45. Where are instructions found for adjustments to a system that is displayed on an electronic display? (AM.II.H.K13)

In the aircraft maintenance manual.

46. Why is grounding and bonding important for aircraft electrical systems? (AM.II.H.K14)

Inadequate bonding or grounding can lead to unreliable operation of systems, EMI, electrostatic discharge damage to sensitive electronics, personnel shock hazard, or damage from lightning strike.

47. What is meant by built-in test equipment (BITE)? (AM.II.H.K15)

BITE is a standard component of maintenance panels that are a part of Electronic Centralized Aircraft Monitor (ECAM) systems. It is standard for monitoring systems to monitor themselves as well as the aircraft systems. All of the system inputs to the flight warning computers can also be tested for continuity from this panel, as well as inputs and outputs of the system data analog converter.

48. What information is allowed to be depicted on electronic displays? (AM.II.H.K16)

All information relating to flight and aircraft systems can be displayed on electronic displays. The main flight instruments are displayed on what is referred to as the primary flight display (PFD) or electronic flight instrument system (EFIS). Data displays for engine parameters, hydraulics, fuel, and other airframe systems are displayed on secondary flight displays or independent displays. These secondary displays are often referred to as multifunction displays (MFDs).

49. What is the purpose of an engine-indicating and crew alerting system (EICAS)? (AM.II.H.K17)

The system monitors aircraft systems for the pilot and displays engine and airframe parameters.

50. What is a head-up display (HUD)? (AM.II.H.K18)

A head-up display proved critical flight information to the pilot on a see-through display that is in the pilot's line of sight.

51. Do the tests described in 14 CFR Part 43, Appendix E, apply to the altimeters installed in all certificated aircraft? (AM.II.H.K19)

No, only those operated under instrument flight rules in controlled airspace.

52. Who is authorized to conduct a static system test specified in 14 CFR §91.411? (AM.II.H.K19)

A certificated Aviation Mechanic holding an Airframe rating.

53. When should a static system leak check be performed? (AM.II.H.K19)

Any time the static system has been opened.

54. **What should be done to the static port that is not being used to conduct the leak test?** (AM.II.H.K19)
It should be taped over in such a way that the tape can not be overlooked or forgotten when the test is completed.

55. **Can an aviation mechanic with an A&P Rating repair an instrument with a cracked glass?** (AM.II.H.K20)
No. The instrument must be replaced.

56. **How much leakage is allowed in the static system of an unpressurized aircraft?** (AM.II.H.K20)
With a pressure differential of 1 inch of mercury, the system must not show a loss of indicated altitude of more than 100 feet in one minute.

57. **How much leakage is allowed in the static system of a pressurized aircraft?** (AM.II.H.K20)
With a pressure differential equal to the maximum cabin differential pressure for which the aircraft is certificated, the system must not show a loss of indicated altitude of more than 2 percent of the equivalent altitude of the maximum cabin differential pressure or 100 feet, whichever is the greater, in one minute.

58. **What indication on the altimeter shows that the pressure inside the static system has been decreased by 1 inch of mercury?** (AM.II.H.K20)
The altimeter will show an increase of approximately 1,000 feet.

59. **What are six tests that must be made when testing an altimeter?** (AM.II.H.K20)
Scale error, hysteresis, after-effect, friction, case leak, and barometric scale error.

60. **What does an angle of attack (AOA) indicator provide for the pilot?** (AM.II.H.K21)
The AOA offers a visual indication of the amount of lift the wing is producing at a given airspeed or angle of bank. The AOA delivers critical information to indicate the actual safety margin above an aerodynamic stall.

61. **What is the purpose of a stick shaker used on some aircraft?** (AM.II.H.K21)
It provides an artificial stall warning to the pilot.

62. **Why is it important for a stall warning system to be adjusted correctly?** (AM.II.H.K21)
A stall warning system is adjusted to allow sufficient margin and warning prior to an actual stall.

63. **What systems does a takeoff warning system monitor?** (AM.II.H.K22)
The takeoff warning system senses when the power lever is advanced and checks for proper position of flaps and trim systems. A warning horn and caution light will be activated if the takeoff configuration is incorrect.

64. **What conditions are indicated by a landing gear warning system?** (AM.II.H.K22)
A red light will illuminate if any of the gear is in an unsafe condition. A green light shows when all landing gear are down and locked. An aural warning will sound a horn if the landing gear are not down when the throttles are retarded for landing.

65. **What is the purpose of bonding straps between flight controls and the attaching structure?** (AM.II.H.K23)

    To provide an electrical path for dissipation of static electricity and to reduce damage from electrostatic discharges and lightning strikes.

66. **Why are instrument panels shock-mounted?** (AM.II.H.K24)

    To absorb low-frequency, high-amplitude shocks.

Risk Management

1. **What risks are associated with using compressed air for cleaning off an instrument panel or the exterior of the aircraft?** (AM.II.H.R1)

   Pressurized air can quickly damage aircraft instruments if allowed to enter pitot/static ports or gyro vacuum lines.

2. **What cautions should be used when washing the exterior of an aircraft?** (AM.II.H.R1)

   Never spray water into pitot/static ports. Water can cause damage or cause the instrument to read incorrectly.

3. **What dangers and challenges arise from reported intermittent warnings or caution annunciator lights?** (AM.II.H.R2)

   An intermittent problem can be difficult to troubleshoot since the system may appear to be working correctly when looking for the problem. However, there was a cause for the intermittent warning or caution and the underlying problem must be understood.

4. **What precautions should be used when working on electrostatic-sensitive equipment?** (AM.II.H.R3)

   Always use a body grounding strap (wrist strap) that connects the mechanic and the aircraft grounding structure. Work benches should be grounded, and components should be stored in anti-static bags.

5. **What precautions are to be used when handling instruments that contain mechanical gyros?** (AM.II.H.R4)

   Do not allow the instruments to be bumped, jarred, or dropped, as damage may occur.

6. **What cautions should be taken when performing a pitot/static system test?** (AM.II.H.R5)

   Use the correct test equipment and procedures. Over-pressurizing the system can cause damage. Upon completion of the leak test, be sure that the system is returned to the normal flight configuration. If it is necessary to block off various portions of a system, check to be sure that all blanking plugs, adaptors, or pieces of adhesive tape have been removed.

Skills

1. **Perform a static system check on an aircraft specified by the examiner. Determine from the proper source if the system meets the requirements for flight under instrument flight rules.** (AM.II.H.S1)

2. **Remove and install instruments as directed by the examiner.** (AM.II.H.S2)
3. **Identify an electric attitude indicator.** (AM.II.H.S2)
4. **Locate a servo-type indicating system.** (AM.II.H.S2)
5. **Install range markings on instrument glass.** (AM.II.H.S3)
6. **Apply instrument glass slippage marks.** (AM.II.H.S3)
7. **Demonstrate to the examiner how to determine barometric pressure using an altimeter.** (AM.II.H.S4)
8. **Demonstrate to the examiner the proper way to adjust the wing flap position sensors to the aircraft manufacturer's specifications.** (AM.II.H.S5)
9. **Determine the proper range markings on an instrument assigned by the examiner.** (AM.II.H.S5)
10. **Locate the instructions for the inspection of a magnetic compass.** (AM.II.H.S6)
11. **Demonstrate to the examiner the correct way to swing a compass and record the results.** (AM.II.H.S6)
12. **Inspect a magnetic compass and explain how to determine if the compass is airworthy.** (AM.II.H.S6)
13. **Identify an aircraft vacuum system.** (AM.II.H.S7)
14. **Demonstrate the correct way to change the air pump in the low-pressure pneumatic system of an aircraft specified by the examiner.** (AM.II.H.S7)
15. **Locate procedure for troubleshooting vacuum operated turn-and-bank instruments.** (AM.II.H.S7)
16. **Locate and explain the troubleshooting procedures for a directional gyro system malfunction.** (AM.II.H.S7)
17. **Demonstrate the correct way to install a sensitive altimeter and check the indication of the altimeter when the barometric scale is set to the local altimeter setting.** (AM.II.H.S8)
18. **Determine the proper altimeter for installation on a given aircraft.** (AM.II.H.S8)
19. **Identify the exhaust gas temperature system components.** (AM.II.H.S9)
20. **Explain how to troubleshoot an electrical resistance thermometer system.** (AM.II.H.S9)

21. **Demonstrate the correct way to change the filter in the gyro instrument system of an aircraft specified by the examiner.** (AM.II.H.S10)

22. **Inspect a vacuum system filter for serviceability.** (AM.II.H.S10)

23. **Adjust the gyro/instrument air pressure.** (AM.II.H.S11)

24. **Check the pitot heater to determine whether it is functioning properly.** (AM.II.H.S11)

25. **Remove and install a heated pitot tube.** (AM.II.H.S11)

26. **Locate the alternate air source on an aircraft.** (AM.II.H.S12)

27. **Inspect an aircraft's alternate air (static) source.** (AM.II.H.S12)

28. **Locate and explain the adjustment procedures for a stall warning system.** (AM.II.H.S13)

29. **Inspect an outside air temperature gauge for condition and operation.** (AM.II.H.S14)

I. Communication and Navigation Systems

References: 14 CFR Part 91; AC 43.13-1, AC 43.13-2; FAA-H-8083-31

Knowledge

1. **Which frequency band is used for long-range communications from an aircraft?** (AM.II.I.K1)
   The high frequency band (2 to 25 megahertz).

2. **What are the two methods of modulating carrier waves for transmitting information?** (AM.II.I.K1)
   Amplitude modulation (AM) and frequency modulation (FM).

3. **What is meant by a transceiver?** (AM.II.I.K2)
   It is a piece of radio communications equipment in which all of the circuits for the receiver and the transmitter are contained in one housing.

4. **Is a certificated airframe mechanic allowed to adjust a communications transmitter?** (AM.II.I.K2)
   No, this requires a license issued by the Federal Communications Commission.

5. **What kind of antenna is used for the ATC transponder?** (AM.II.I.K3)
   A UHF stub antenna.

6. **What is the preferred location for the ATC transponder antenna?** (AM.II.I.K3)
   On the center line of the belly of the aircraft as far from any other antenna as is practical.

7. **What are the three basic types of antennas used in aviation?** (AM.II.I.K3)
   Dipole antenna, Marconi antenna, and loop antenna.

8. **What type of antenna are most aircraft VHF communication antennas?** (AM.II.I.K3)
   Marconi antennas.

9. **How do static wicks aid the aircraft communication radios?** (AM.II.I.K3)
   Static wicks aid in the dissipation of static that builds up on the aircraft. Built-up static can cause radio receiver interference caused by corona discharge emitted from the aircraft as a result of precipitation static.

10. **What is the purpose of an intercom system?** (AM.II.I.K4)
    To aid in communication between the flight crew, or between the flight crew and the cabin crew.

11. **What kind of conductor is used to connect a VHF or UHF antenna to the receiver or transmitter?** (AM.II.I.K5)
    Coaxial cable.

12. **What is a coaxial cable?** (AM.II.I.K5)
    A type of two-conductor electrical cable in which the center conductor is encased in insulation inside a braided shield that serves as the outer conductor. Coaxial cables are normally used for attaching radio receivers and transmitters to antennas.

13. **How can you determine the proper coaxial cable and connectors to use in an aircraft radio installation?** (AM.II.I.K5)
    Refer to the radio installation instructions for the correct part number for the cable and connectors.

14. **What precautions should be taken when installing coaxial cable between a radio transmitter and its antenna?** (AM.II.I.K5)
    The routing should be as direct as possible, there should be no sharp bends in the coax, and it should be kept away from heat that could soften the insulation.

15. **What is a BNC connector?** (AM.II.I.K5)
    A coaxial cable connector that is connected by inserting the guide pins on the male connector into slots in the female connector and twisting the connector one quarter turn.

16. **What kind of antenna is used for VHF communications?** (AM.II.I.K5)
    A vertically polarized whip antenna.

17. **Which frequency band is used for most aircraft communications?** (AM.II.I.K5)
    The VHF band, between 30 and 300 megahertz.

18. **What is the Aircraft Communications Addressing and Reporting System (ACARS), and how does it work?** (AM.II.I.K6)

    ACARS is a two-way communication link between an airliner in flight and the airline's main ground facilities. Data is collected in the aircraft by digital sensors and is transmitted to the ground facilities. Replies from the ground may be printed out so the appropriate flight crewmember can have a hard copy of the response.

19. **On what two frequencies does the emergency locator transmitter operate?** (AM.II.I.K7)

    121.5 and 243.0 megahertz.

20. **Where is the ELT transmitter normally located on an aircraft?** (AM.II.I.K7)

    In the tail of the aircraft or as far aft as possible, so it will be least likely to be damaged in a crash.

21. **How often should ELT batteries be replaced or recharged?** (AM.II.I.K7)

    When the transmitter has been in use for more than 1 cumulative hour, when the battery expiration date has been reached, or when 50% of their useful life or charge has expired.

22. **How can you know when an ELT battery must be replaced or recharged?** (AM.II.I.K7)

    By the date marked on the outside of the transmitter.

23. **How often must ELTs be inspected for proper installation, battery corrosion, operation of the controls and sensor, and the presence of the radiated signal?** (AM.II.I.K7)

    Every 12 calendar months.

24. **What causes an ELT to actuate?** (AM.II.I.K7)

    An inertia switch that detects an impact parallel to the longitudinal axis of the aircraft as would occur in a crash.

25. **How is an ELT tested to determine that it is working?** (AM.II.I.K7)

    Actuate the test switch and listen on 121.5 or 243.0 MHz. Make the test during the first five minutes of the hour and do not allow the ELT to operate for more than 3 sweeps. If the ELT is operated outside of this time frame, you should contact the control tower before conducting the test.

26. **What two types of antenna are used with most ADF receivers?** (AM.II.I.K8)

    A directional loop antenna and a nondirectional sense antenna.

27. **Why is it necessary to install a doubler on the inside of the aircraft skin when antenna is mounted on the skin?** (AM.II.I.K8)

    The doubler reinforces the skin so wind loads on the antenna will not cause the skin to flex and crack.

28. **What is the preferred location for a VOR antenna on an airplane?** (AM.II.I.K9)

    On top of the aircraft, along the center line of the fuselage.

29. **Which component of the Instrument Landing System shares the antenna with the VOR?** (AM.II.I.K9)
The ILS localizer.

30. **In which frequency band does the VOR equipment operate?** (AM.II.I.K9)
In the VHF band, between 108.0 and 117.95 megahertz.

31. **What is the preferred location for a DME antenna?** (AM.II.I.K10)
Along the center line of the belly of the aircraft, as far from any other antenna as is practical.

32. **In what frequency band does the DME equipment operate?** (AM.II.I.K10)
In the UHF band, between 962 and 1,024 megahertz, and between 1,151 and 1,213 megahertz.

33. **What does a DME system measure?** (AM.II.I.K10)
Direct distance to the station that the signal is received from.

34. **What two signals are transmitted from an ILS ground-based station and then received in the aircraft?** (AM.II.I.K11)
A localizer signal provides horizontal guidance to the centerline of the runway. A separate glideslope provides vertical guidance of the aircraft down the proper slope to the touchdown point on the runway.

35. **Where do GPS units receive their signal from?** (AM.II.I.K12)
From satellites that orbit around the Earth.

36. **How many satellites are typically in view of a GPS receiver?** (AM.II.I.K12)
Between five and eight satellites.

37. **How does a GPS receiver determine its position?** (AM.II.I.K12)
The amount of time it takes for signals to reach the aircraft GPS receiver from transmitting satellites is combined with each satellite's exact location to calculate the position of an aircraft.

38. **What aircraft system does TCAS (traffic collision avoidance system) use for air-to-air monitoring and alerting?** (AM.II.I.K13)
The aircraft transponder.

39. **What is a TCAS Resolution Advisory (RA)?** (AM.II.I.K13)
This is an aural command to the pilot to take a specific evasive action (i.e., DESCEND). The computer is programmed such that the pilot in the encroaching aircraft receives an RA for evasive action in the opposite direction (if it is TCAS II equipped).

40. **What type of weather is shown by an onboard weather radar system?** (AM.II.I.K14)
Onboard weather radar signals bounce off of precipitation only. Clouds do not create a return.

41. **Weather aids referred to as weather radar cover what three types of systems?** (AM.II.I.K14)
Onboard weather radar, lightning detectors, and satellite or other source weather radar information that comes from an outside source.

42. **What is the primary advisory that a radio altimeter is used for?** (AM.II.I.K15)
The landing decision height when on an instrument approach.

43. **Where does a ground proximity warning system receive its information from?** (AM.II.I.K15)
From a radio altimeter.

44. **What is the basic purpose of an autopilot?** (AM.II.I.K16)
It frees the human pilot from continuously having to fly the aircraft, and flies with a high degree of precision. It also couples with various electronic navigational aids.

45. **What are the basic subsystems of an automatic flight control system?** (AM.II.I.K16)
Command, error-sensing, correction, and follow-up.

46. **What type of device is normally used in the error-sensing subsystem?** (AM.II.I.K16)
Gyros.

47. **What are three types of servos that are used in the correction subsystem?** (AM.II.I.K16)
Hydraulic, pneumatic, and electric.

48. **What is the purpose of the follow-up subsystem in an autopilot?** (AM.II.I.K16)
It stops the control movement when the surface has deflected the proper amount for the signal sent by the error sensor.

49. **What are the two parameters that auto-throttles can maintain or attain, and how do they function?** (AM.II.I.K17)
The two parameters are speed mode and thrust mode. In speed mode, the throttle is positioned to maintain a set target speed. In thrust mode, the engines are maintained at a fixed power setting to meet the different phases of flight (takeoff, climb, descent, etc.).

50. **What are the two types of helicopter stability augmentation systems?** (AM.II.I.K18)
A force trim system holds the cyclic control in the position at which it was released. A more advanced system uses electric actuators that make inputs to the hydraulic servos based on control commands from a computer that senses helicopter attitude.

51. **At what altitudes does a radio altimeter begin to measure aircraft height above the ground?** (AM.II.I.K19)
Below 2,500 feet.

52. **What data does an ADS-B Out system transmit?** (AM.II.I.K20)
Aircraft GPS location, altitude, velocity, and speed.

53. **Where does an ADS-B Out system transmit its data to?** (AM.II.I.K20)
Ground-based stations and other aircraft.

54. **In addition to information on other aircraft, what other information is available on ADS-B In?** (AM.II.I.K20)
Weather text and graphics information, ATIS information, and NOTAMs.

55. **What is the purpose of a transponder?** (AM.II.I.K21)
A transponder provides positive identification and location of an aircraft on the radar screens of ATC.

56. **How does an altitude encoder interface with a transponder?** (AM.II.I.K21)
For aircraft equipped with an altitude encoder, the transponder also provides the pressure altitude of the aircraft to be displayed adjacent to the ATC on-screen blip that represents the aircraft.

Risk Management

1. **Why is it important to follow the recommended procedures for testing ELTs?** (AM.II.I.R1)
A signal from an ELT can trigger a search and rescue operation, or at the very least, an inquiry to find out where the signal is coming from. Testing inside a metal hangar does not prevent the ELT's signal from being picked up by Cospas-Sarsat satellite system.

2. **What risks are associated with performing maintenance on high power/high frequency systems such as weather radar?** (AM.II.I.R2)
Physical harm is possible from the high energy radiation that is emitted.

3. **What precautions should be considered when routing wire harnesses?** (AM.II.I.R3)
Wire harnesses need to be protected from chafing and from interfering with other systems, especially those that have movement (retractable landing gear, flight controls, etc.). The routing of wire bundles also needs to consider the environmental condition that will be encountered (heat, corrosive environments, moisture, etc.). Wire bundles also generate magnetic fields and should not be routed close to other equipment that is sensitive to magnetic interference.

4. **What risks are associated with mounting aircraft antennas?** (AM.II.I.R4)
Antennas must be able to withstand high air loads and require a mounting doubler to be installed on the aircraft skin. The location of the antenna must not interfere with other locations and must be mounted in an area where signals can be optimally transmitted and received. Antennas must be electrically matched to the receiver and transmitter that they serve.

5. **Why is it important for static discharge wicks to be mounted on the trailing edges of control surfaces, wing tips, and the vertical stabilizer?** (AM.II.I.R5)
It is important that static wicks discharge static at points a critical distance away from avionics antennas where there is little or no coupling of the static to cause interference or noise in the radios.

6. **What risks are associated with working on live electrical systems?** (AM.II.I.R6)

   The risks of working on live electrical circuits include inadvertent shorting of a positive voltage to ground, which could cause damage to a component, or inadvertent activation of systems. Additionally, power systems can cause electrical shock and injury to personnel.

Skills

1. **Make a required list of placards for communication and navigation equipment.** (AM.II.I.S1)
2. **Demonstrate to the examiner the correct way to check the proper operation of an autopilot.** (AM.II.I.S2)
3. **Demonstrate to the examiner where to find autopilot inspection procedures.** (AM.II.I.S2)
4. **Locate and identify to the examiner the servos used in an autopilot system.** (AM.II.I.S3)
5. **Locate and identify the antenna for the VOR, the DME, the ATC transponder, the ADF, and the glide slope, and the VHF communications equipment.** (AM.II.I.S4)
6. **Demonstrate to the examiner the correct way to install and check the operation of a VHF nav/com transceiver.** (AM.II.I.S5)
7. **Correctly install a BNC connector on a piece of coaxial cable.** (AM.II.I.S6)
8. **Inspect a coaxial cable for security.** (AM.II.I.S6)
9. **Identify the transponder transmission line.** (AM.II.I.S6)
10. **Demonstrate to the examiner the correct way to check the ELT for operation.** (AM.II.I.S7)
11. **Demonstrate to the examiner the correct way to check the status of the ELT batteries, and to replace or recharge them as is appropriate.** (AM.II.I.S8)
12. **Demonstrate to the examiner the way to determine that the ELT is not transmitting when the aircraft is being shut down.** (AM.II.I.S8)
13. **Inspect the electronic equipment mounting base for security and condition.** (AM.II.I.S9)
14. **Inspect the electronic equipment shock-mounting bonding jumpers for security and resistance.** (AM.II.I.S10)
15. **Inspect the static discharge wicks for security and resistance.** (AM.II.I.S11)

16. **Make the proper entry in the aircraft records for the installation of the transceiver. Include the revised weight and balance record.** (AM.II.I.S12)

17. **Inspect a radio installation for condition and security.** (AM.II.I.S12)

18. **Locate the installation procedures for antennas including mounting and coaxial connections.** (AM.II.I.S13)

J. Aircraft Fuel Systems

References: AC 43.13-1; FAA-H-8083-31

Knowledge

1. **What are two types of fuel cells used in modern aircraft?** (AM.II.J.K1)
   Integral fuel cells (cells that are a sealed-off portion of the structure), and bladder-type cells.

2. **What types of operations are fuel systems designed and certified for?** (AM.II.J.K1)
   Any maneuver or operating condition for which the aircraft is certified.

3. **Where can you find the correct part number for the fuel quantity sensor to be installed in an aircraft fuel tank?** (AM.II.J.K2)
   In the Illustrated Parts Catalog for the aircraft.

4. **Where are fuel system strainers located?** (AM.II.J.K2)
   One strainer is located in the outlet to the tank, and the main strainer is located in the fuel line between the outlet of the fuel tank and the inlet to the fuel metering device.

5. **Why do the fuel strainers used on some turbine-engine aircraft have a warning device that signals the flight crew when the strainer is beginning to clog up?** (AM.II.J.K2)
   Strainers clog because ice forms on the filter element. The flight crew can route the fuel through a fuel heater to melt the ice.

6. **How does water appear in the fuel drained from the tank sumps?** (AM.II.J.K2)
   Water appears as a clear globule in the bottom of the container used to collect the fuel drained from the sumps.

7. **Why must an aircraft fuel valve have a detent in its operating mechanism?** (AM.II.J.K2)
   The detent gives the pilot a positive indication by feel when the selector valve is in the full ON and full OFF position.

8. **What must be done before a fuel selector valve can be removed from an aircraft?** (AM.II.J.K2)

   The tank to which the valve is connected must be drained or otherwise isolated.

9. **What must be done after a fuel valve is replaced?** (AM.II.J.K2)

   The fuel system must be checked for leaks.

10. **Why are fuel tanks divided into compartments or have baffles installed in them?** (AM.II.J.K3)

    The compartments or baffles keep the fuel from surging back and forth as the aircraft changes its attitude in flight.

11. **How is a fuel leak indicated on a reciprocating-engine-powered aircraft?** (AM.II.J.K3)

    The dye in the gasoline stains the area around the leak.

12. **What is the function of the sump in a fuel tank?** (AM.II.J.K3)

    The sump, or lowest point, in a fuel tank is located below the fuel outlet. The sump collects water and contaminants and will have a drain to remove them.

13. **What are the three different types of fuel tanks?** (AM.II.J.K3)

    Built-up tanks, integral tanks, and bladder tanks.

14. **What problem can occur in a bladder tank with a blocked vent?** (AM.II.J.K3)

    As internal tank pressure decreases, the pressure differential between the atmosphere and the tank can cause the tank restraints to fail and the tank can collapse and block the fuel outlet.

15. **How do you prepare an aluminum fuel tank for welding?** (AM.II.J.K3)

    Remove all fuel and thoroughly wash the tank, inside and out, with hot water and detergent. The tank should be purged with live steam for at least 30 minutes to remove all residual fuel and vapors.

16. **After welding an aluminum fuel tank, to what pressure should the tank be tested?** (AM.II.J.K3)

    3.5 psi.

17. **To what fuel flow is an aircraft fuel system designed for?** (AM.II.J.K4)

    Each fuel system must be constructed and arranged to ensure fuel flow at a rate and pressure established for proper engine and auxiliary power unit (APU) functioning under each likely operating condition.

18. **What markings must appear near the filler opening of the fuel tanks on a reciprocating-engine-powered aircraft, and on a turbine-powered aircraft?** (AM.II.J.K5)

    On a reciprocating engine powered-aircraft: the word "Avgas" and the minimum grade of fuel. On a turbine engine powered aircraft: the words "Jet Fuel," the permissible fuel designations, the maximum permissible fueling supply pressure, and the maximum permissible defueling pressure.

19. **Why is it important to drain all of the fuel sumps before the first flight of the day?** (AM.II.J.K5)

    Water can condense in the fuel tanks and it must be drained out before the aircraft is safe for flight.

20. **Is the procedure of draining the sumps with the aircraft in the ground attitude an assurance that all of the water is removed from the tanks?** (AM.II.J.K5)

    No; some aircraft require special procedures to remove all of the water. Refer to the airplane flight manual or pilot's operating handbook.

21. **Why is a fuel jettisoning or dump system required for some aircraft?** (AM.II.J.K6)

    A fuel dump system is installed on most large aircraft and allows the flight crew to jettison, or dump, fuel to lower the gross weight of the aircraft to its allowable landing weight.

22. **What color is grade 100-LL fuel?** (AM.II.J.K7)

    Blue.

23. **How can you detect jet fuel in the sample taken from the tank sump drains?** (AM.II.J.K7)

    Jet fuel is basically clear and it has an odor similar to kerosine.

24. **What must be done after a fuel system strainer has been cleaned or replaced?** (AM.II.J.K8)

    The system must be tested for leaks by pressurizing the system with the boost pump, if one is used.

25. **What must be done if the sump drains sample shows traces of jet fuel?** (AM.II.J.K8)

    The fuel system must be drained and flushed out with the proper grade of aviation gasoline.

26. **What is used as the sensor in the fuel tank for an electronic-type fuel quantity indicating system?** (AM.II.J.K9)

    Tubular capacitors that extend across the fuel tank from top to bottom.

27. **What is the principle upon which the electronic-type fuel quantity indicating system operates?** (AM.II.J.K9)

    Tubular capacitors extending across the fuel tanks change their capacitance as the fuel level changes. The dielectric constant (k) of the fuel is approximately twice that of air.

28. **What is used as a sensor in the fuel tank for the older resistance-type fuel quantity indicating system?** (AM.II.J.K9)

    A variable resistor with an arm that is moved by a float riding on top of the fuel in the tank.

29. **What is the purpose of a drip gauge in the fuel tank of a large aircraft?** (AM.II.J.K9)

    The drip gauge allows a mechanic to check the fuel level in a tank from the bottom of the tank.

30. **When must the fuel quantity indicating system indicate "zero?"** (AM.II.J.K9)
During level flight when the fuel in the tank is equal to the unusable fuel supply.

Risk Management

1. **What risks are associated with fuel system maintenance?** (AM.II.J.R1)
Improper fuel system maintenance can lead to failure of the fuel system or inflight fires. Always follow the manufacturer's procedures and after completion, have a second mechanic check fittings for security and proper routing of fuel lines.

2. **What risks are associated with fuel contamination?** (AM.II.J.R2)
Fuel contamination can lead to engine failure. Verify that fuel is of the proper type and check fuel sumps for water and other contaminations.

3. **What risks are associated with fuel spills?** (AM.II.J.R3)
Vapors coming from spilled fuel can ignite, causing a fire. Fuel spills should be addressed by immediate action that corresponds to the type of fuel and the amount of fuel spilled. Significant fuel spills should be handled by the airport fire department.

4. **What safety procedures should be followed when entering a large fuel tank?** (AM.II.J.R4)
All fuel must be emptied from the tank and strict safety procedures must be followed. Fuel vapors must be purged from the tank and respiratory equipment must be used by the aviation mechanic. A full-time spotter must be positioned just outside of the tank to assist if needed.

5. **What precautions should be used when defueling an aircraft?** (AM.II.J.R5)
Follow aircraft manufacturer's procedures and use the proper equipment. On large aircraft, the same fueling port can be used for defueling.

Skills

1. **Inspect the fuel selector valves of an aircraft and determine whether there is positive indication of the valve being fully on and fully off.** (AM.II.J.S1)

2. **Demonstrate to the examiner the correct way to remove and replace a fuel selector valve.** (AM.II.J.S1)

3. **Inspect a metal fuel tank.** (AM.II.J.S2)

4. **Inspect a bladder fuel tank.** (AM.II.J.S2)

5. **Inspect an integral fuel tank.** (AM.II.J.S2)

6. **Troubleshoot a fuel system provided by the examiner. Determine any faults in the system and repair or replace the failed component.** (AM.II.J.S3)

7. **Inspect a fuel selector valve provided by the examiner.** (AM.II.J.S4)

8. **Inspect a fuel selector valve and demonstrate to the examiner how to determine if the valve is operating correctly.** (AM.II.J.S5)

9. **Troubleshoot the supplied selector valve system.** (AM.II.J.S6)

10. **Demonstrate to the examiner the correct way to sample the fuel taken from tank sump drains. Explain the way to detect water and rust contamination, and contamination with jet fuel.** (AM.II.J.S7)

11. **Demonstrate to the examiner the way to perform the following operations on the main fuel strainer of the aircraft specified by the examiner:**
    a. **Remove and properly clean the strainer element.**
    b. **Reinstall the strainer element and check the filter for leaks.**
    c. **Safety the strainer by the method specified by the aircraft manufacturer.**

    (AM.II.J.S8)

12. **Locate the fuel system inspection instructions.** (AM.II.J.S8)

13. **Troubleshoot the fuel system.** (AM.II.J.S8)

14. **Inspect a remote fuel indicating system.** (AM.II.J.S9)

15. **Demonstrate to the examiner the way to perform the following operations on an electronic-type fuel quantity indicating system:**
    a. **Locate the correct part number for the fuel tank probe specified by the examiner.**
    b. **Install the probe in the fuel tank.**
    c. **Calibrate the system in accordance with instructions given by the aircraft manufacturer.**
    d. **Make the appropriate maintenance record for the replacement of the fuel tank probe.**

    (AM.II.J.S9)

16. **Demonstrate to the examiner the correct way to use a fuel drip gauge to measure fuel quantity.** (AM.II.J.S9)

17. **Locate the fuel system's operating instructions.** (AM.II.J.S10)

18. **Locate in the proper documentation the correct grade of fuel to be used in an aircraft specified by the examiner.** (AM.II.J.S10)

19. **Locate the fuel system's inspection instructions.** (AM.II.J.S11)

20. **Locate the fuel system's crossfeed procedures.** (AM.II.J.S12)

21. **Locate the fuel system's required placards.** (AM.II.J.S13)

22. **Using a POH furnished by the examiner, identify the minimum grade of fuel allowed for the aircraft.** (AM.II.J.S13)

23. **Locate the fuel system's defueling procedures.** (AM.II.J.S14)

24. **Troubleshoot the fuel pressure warning system.** (AM.II.J.S15)

25. **Locate the troubleshooting procedures for fuel temperature systems.** (AM.II.J.S16)

26. **Remove and/or install a fuel quantity transmitter.** (AM.II.J.S17)

27. **Troubleshoot a fuel quantity indicating system.** (AM.II.J.S18)

K. Aircraft Electrical Systems

References: AC 43.13-1; FAA-H-8083-31

Knowledge

1. **What size generator must be used in an aircraft electrical system if the connected electrical load is 30 amps, and there is no way of monitoring the generator output?** (AM.II.K.K1)
   When monitoring is not practical, the total continuously connected electrical load must be no more than 80% of the rated generator output. This would require a generator with a rating of 37.5 amps. Practically, a 40-amp generator would be installed.

2. **What is meant by "flashing" the field of a generator?** (AM.II.K.K1)
   Restoring the residual magnetism to the frame of the generator. This is done by passing battery current through the field coils in the direction it normally flows when the generator is operating.

3. **What is meant by paralleling the generators in a multi-engine aircraft?** (AM.II.K.K1)
   Adjusting the voltage regulators so all the generators share the electrical load equally.

4. **What are three types of voltage regulators used with aircraft generators?** (AM.II.K.K1)
   Vibrator-type, carbon-pile type, and solid state-type.

5. **How many electrical phases are in aircraft alternators?** (AM.II.K.K2)
   Three.

6. **What is the advantage of using AC power over DC power?** (AM.II.K.K2)
   AC voltages can be stepped up to higher voltages, which means that current goes down and smaller wires can be used to save weight. The voltage is stepped back down at the point of use.

7. **What is the frequency of AC voltage used in aircraft?** (AM.II.K.K2)
   400 Hz.

8. **What is a starter-generator?** (AM.II.K.K3)
A starter-generator is a generator and a starter combined into a single unit.

9. **How many sets of field windings are contained in a starter-generator?** (AM.II.K.K3)
There are two sets of field windings. One field is used to start the engine and one is used for the generation of electrical power.

10. **How does a constant speed drive (CSD) unit maintain a constant output frequency of the AC generator?** (AM.II.K.K4)
By maintaining a constant speed of the generator as the engine speed varies.

11. **What is an integrated drive generator (IDG)?** (AM.II.K.K4)
When a constant speed drive (CSD) is combined within the alternator housing, the assembly is known as an IDG.

12. **What is meant by a trip-free circuit breaker?** (AM.II.K.K5)
A circuit breaker that cannot be closed while a fault exists, regardless of the position of the operating handle.

13. **What is meant by a slow-blow fuse?** (AM.II.K.K5)
A fuse that will allow more current than its rating to flow for a short period of time, but will open the circuit if more than its rated current continues to flow.

14. **What is the function of a fuse or circuit breaker in an aircraft electrical circuit?** (AM.II.K.K5)
It protects the wiring from an excess of current. It will open the circuit if enough current flows to heat the wire.

15. **What are two principles upon which circuit breakers operate?** (AM.II.K.K5)
Magnetic circuit breakers open a circuit when the current creates a strong enough magnetic field. Thermal circuit breakers open a circuit when the current causes enough heat.

16. **What circuit in an aircraft electrical system is not required to have a circuit protective device?** (AM.II.K.K5)
The main circuit for starter motors, used during starting only.

17. **Is an automatic-reset circuit breaker approved for aircraft electrical circuits?** (AM.II.K.K5)
No, a manual operation is needed to restore service after the circuit breaker has tripped.

18. **How does a vibrator-type voltage regulator maintain a constant voltage?** (AM.II.K.K5)
When the voltage rises above the desired value, an electromagnetic relay opens and inserts a resistor in the generator field circuit, decreasing the field current and lowering the generator output voltage.

19. **What two components are normally housed with a vibrator voltage regulator in a single-unit generator control?** (AM.II.K.K5)
A current limiter and a reverse-current cutout relay.

20. **What are the two types of inverters found in aircraft?** (AM.II.K.K6)
Rotary inverters and static inverters.

21. **What is the purpose of an inverter?** (AM.II.K.K6)
To convert DC electricity to AC electricity.

22. **How does a rotary inverter create AC voltage?** (AM.II.K.K6)
By using a DC motor to spin an AC generator.

23. **Where could you find the part number of a switch in an aircraft electrical system?** (AM.II.K.K7)
In the equipment table or bill of materials on the electrical circuit diagram for the aircraft.

24. **What is the main disadvantage of aluminum wire over copper wire for use in an aircraft electrical system?** (AM.II.K.K7)
Aluminum wire is more brittle than copper. It is more subject to breakage when it is nicked or when it is subjected to vibration.

25. **What size aluminum wire would be proper to replace a piece of four-gauge copper wire?** (AM.II.K.K7)
Two-gauge. When you substitute aluminum wire for copper wire, use a wire that is two gauge numbers larger.

26. **What is the smallest size aluminum wire that is approved for use in aircraft electrical systems?** (AM.II.K.K7)
Six-gauge.

27. **Can you substitute aluminum wire for copper wire?** (AM.II.K.K7)
Yes. However, you must use aluminum wire two wire gauge sizes larger to carry the same current. Aluminum wire should never be used in runs of three feet or less, or in communication and navigation systems.

28. **Can wire in free air carry the same current as wire in a bundle?** (AM.II.K.K7)
Wire in free air may be used to a higher current level than the same size wire in a bundle. Always refer to an electrical wire size selection chart.

29. **What are common causes of wire failure in a crimped connector?** (AM.II.K.K7)
The wire was not inserted far enough into the connector and/or the connector was excessively crimped.

30. **What is one of the first things to check if an electrical component does not operate?** (AM.II.K.K7)
Always start with simple solutions and move to more complex. Is the component turned on? Is power available? For example, if a light doesn't function is the system turned on, is the aircraft electrical system energized, or is the bulb burnt out?

31. **What may be provided by the manufacturer to assist in troubleshooting?** (AM.II.K.K7)
Manufacturers often provide a troubleshooting logic chart.

32. **What are the four basic steps of troubleshooting?** (AM.II.K.K7)
Know how the system should operate. Observe the way the system is operating. Divide the system into smaller segments to isolate trouble. Look for the obvious problem first.

33. **When routing a fluid line parallel to an electrical wire bundle, which should be on top?** (AM.II.K.K7)
The electrical wire bundle should be on top.

34. **Why must a switch be derated if it is used in a circuit that supplies incandescent lamps?** (AM.II.K.K8)
The high-inrush current caused by the low resistance of the cold filaments requires that the switches be derated.

35. **What is the purpose of aircraft wire shielding?** (AM.II.K.K9)
To eliminate electromagnetic interference by applying a metallic covering to the wiring.

36. **In aircraft that are of aluminum construction, what is one of the primary design elements for lightning protection?** (AM.II.K.K10)
The use of bonding or grounding straps between all major components and flight controls on the airplane.

37. **How are composite aircraft protected from the damage of lightning strikes?** (AM.II.K.K10)
By using a conductive copper for aluminum mesh in the outer layers of the composite structure.

38. **What is the purpose of grounding straps on an instrument panel?** (AM.II.K.K11)
To ensure electrical continuity from the panel to the airframe for proper system operation and to reduce the potential for EMI.

39. **What are the colors and positions of aircraft position lights?** (AM.II.K.K12)
Left side/wing tip—red
Right side/wing tip—green
Aft—white

40. **What is the purpose of wing inspection lights?** (AM.II.K.K12)
To permit visual detection of ice formation on wing leading edges while flying at night.

41. **What two things must you take into consideration when selecting the wire size to use in an aircraft electrical system installation?** (AM.II.K.K13)
The current carrying capability of the wire and the voltage drop caused by the current flowing through the wire.

42. **What is the maximum number of wires that should be connected to any single stud in a terminal strip?** (AM.II.K.K13)
Four.

43. How is a wire bundle protected from chafing where the bundle goes through a hole in a fuselage frame or bulkhead? (AM.II.K.K13)

The edges of the hole are covered with a flexible grommet, and the bundle is secured to the structure with a cushioned clamp.

44. What kind of clamp is used to secure a wire bundle to the aircraft structure? (AM.II.K.K13)

A cushioned clamp.

45. What is the composition of 60/40 solder? (AM.II.K.K14)

60% tin and 40% lead.

46. What is the difference between soldering and brazing? (AM.II.K.K14)

Soldering is a method of joining metal parts with a molten nonferrous alloy that melts at a temperature below 800°F. Brazing is essentially the same except the brazing alloy melts at a temperature higher than 800°F but lower than the melting temperature of the metal on which it is used.

47. Why must acid-core solder never be used on electrical wire? (AM.II.K.K14)

The acid causes the wire to corrode.

48. What determines the strength of a soldered joint? (AM.II.K.K14)

The mechanical connection of the joint, not the solder.

49. Why are solderless splices usually better than soldered splices in the wiring of an aircraft electrical system? (AM.II.K.K14)

Soldered joints are usually stiff, and vibration can harden the wire and cause it to break. Solderless splices are designed to keep the joint flexible so vibration cannot cause the wire to break.

50. What is an SPDT switch? (AM.II.K.K15)

A single-pole, double-throw switch.

51. If no specific instructions are available, which way should the operating handle of an electrical switch move to turn a component on? (AM.II.K.K15)

Forward or upward.

52. What color insulator on a preinsulated solderless connector indicates that the connector is proper for a 10-gauge wire? (AM.II.K.K15)

Yellow.

53. How many splices are allowed in a single wire run? (AM.II.K.K15)

One.

54. When checking an electrical circuit for proper operation, what are the most efficient measurements (voltage, current, or resistance)? (AM.II.K.K16)

Voltage measurements are usually the quickest way to verify the completeness of the circuit by measuring the voltage at various points through the circuit.

55. **What does the specific gravity of the electrolyte of a lead-acid aircraft battery indicate?** (AM.II.K.K17)

    The amount of acid relative to the water in the electrolyte. This is an indication of the state of charge of the battery.

56. **What are the most likely causes of a battery with a distorted or swollen battery case?** (AM.II.K.K17)

    The battery has been overcharged or over-discharged, causing overheating, or it could be caused by a plugged vent cap.

57. **On NiCd batteries, what is the minimum cell voltage at the end of the charge cycle?** (AM.II.K.K17)

    1.55 volts.

58. **How is the state of charge determined on a NiCd battery?** (AM.II.K.K17)

    The only accurate way to determine the state of charge of a NiCd battery is by a measured discharge with a NiCd battery charger and following the manufacturer's instructions.

Risk Management

1. **What are the risks associated with testing and troubleshooting electrical systems or components?** (AM.II.K.R1)

   Incorrect procedures during testing and troubleshooting can lead to damage of components and systems, or even personal injury. Always follow the manufacturer's procedures for testing and troubleshooting.

2. **What are the risks associated with connecting or disconnecting external power?** (AM.II.K.R2)

   There is a risk that system switches or controls have been moved while the power is off. Connecting power could cause the systems to move. Before connecting power, confirm the position of landing gear controls and flap position levers.

3. **What care should be taken when working on energized circuits or systems?** (AM.II.K.R3)

   An energized system may begin to operate or move unexpectedly while working on the system. Follow the maintenance manual procedures for adjusting or servicing energized circuits and systems.

4. **What risks are associated with performing maintenance in areas containing wires and wire bundles?** (AM.II.K.R4)

   If the bundles are exposed, it may be easy to inadvertently catch on a wire or bundle with a tool or support equipment, such as step ladders. Take care to avoid and protect wires and wire bundles when working in these areas.

5. **What risks are associated with routing and securing wires and wire bundles?** (AM.II.K.R5)

   Misrouted wires can interfere with other systems or become damaged by surrounding structure and components. To mitigate these risks, route wires and wire bundles with sufficient clearance from any moving components and any structure where chafing could occur. Use proper standoffs and cushioned clamps to maintain clearance and chafe protection.

6. **What are the risks of using the incorrect wire size in a circuit?** (AM.II.K.R6)

   A wire size that is too small will not be able to handle the current and may melt or catch fire. Correct usage of wire sizing charts is critical when determining wire sizes.

7. **When replacing a wire terminal, why is it important to select the correct size and follow proper installation procedures?** (AM.II.K.R7)

   Incorrect terminals or poor installation can cause the connection to fail. Always follow terminal size and installation procedures, including the use of approved or recommended tools for the installation.

8. **What are the risks associated with soldering wire connections?** (AM.II.K.R8)

   Soldered connections are stiff and can become brittle, causing premature failure. When required, soldered connections must be done correctly and supported to prevent movement of the wires.

9. **What are the risks associated with incorrect soldering practices?** (AM.II.K.R9)

   A cold solder joint will result in poor connection and early failure. Excessive heat during soldering can damage other wires and components. Excessive solder can cause bridging to other wires or terminals or too much solder in the wire is prone to becoming brittle. Follow professional soldering practices to prevent premature failure of the solder connection.

Skills

1. **Identify the flight deck lighting circuits.** (AM.II.K.S1)

2. **Test and troubleshoot the electrical system supplied by the examiner.** (AM.II.K.S1)

3. **Selecting the proper terminal and tools, install a solderless terminal on a piece of electrical wire furnished by the examiner.** (AM.II.K.S2)

4. **Splice an electrical wire, using the correct type of splice and the correct insulation.** (AM.II.K.S2)

5. **Correctly attach wires to the terminals of a quick-disconnect connector.** (AM.II.K.S3)

6. **Use a wire diagram provided by the examiner to identify components in the supplied circuit.** (AM.II.K.S4)

7. **Demonstrate to the examiner the correct way to make a soft-solder connection between two stranded copper wires.** (AM.II.K.S5)

8. **Demonstrate to the examiner the correct way to solder a stranded copper wire into a quick-disconnect connector.** (AM.II.K.S5)

9. **Troubleshoot an electrical circuit with an open or short circuit.** (AM.II.K.S6)

10. **Demonstrate to the examiner how to troubleshoot the supplied airframe electrical circuit.** (AM.II.K.S6)

11. **Explain to the examiner the correct way to adjust the voltage controlled by a vibrator-type voltage regulator.** (AM.II.K.S6)

12. **Explain to the examiner the way a carbon-pile voltage regulator operates and the way the paralleling circuit allows the generators to share the load equally.** (AM.II.K.S6)

13. **Select the proper switch to replace one specified by the examiner, install it, check its operation, and record the replacement in the aircraft maintenance records.** (AM.II.K.S7)

14. **Based on a given electrical circuit, select and install the appropriate switches.** (AM.II.K.S7)

15. **Select and install the correct type of wiring in an electrical circuit.** (AM.II.K.S7)

16. **Correctly install bonding jumpers.** (AM.II.K.S7)

17. **Select the proper circuit breaker to replace one specified by the examiner, install it, check its operation, and record the replacement in the aircraft maintenance records.** (AM.II.K.S7)

18. **Based on a given electrical system, select and install fuses and/or circuit breakers.** (AM.II.K.S7)

19. **Secure an electrical wire bundle to an aircraft structure using the proper clamps and grommets.** (AM.II.K.S8)

20. **Demonstrate to the examiner the correct way to tie an electrical wire bundle with spot ties.** (AM.II.K.S8)

21. **Determine the electrical load in a given aircraft system.** (AM.II.K.S9)

22. **Manufacture and install bonding jumpers on the system provided by the examiner.** (AM.II.K.S10)

23. **Given the specifications of an aircraft generator, find its rated current output.** (AM.II.K.S11)

24. **Explain to the examiner the correct way to flash the field of an aircraft generator.** (AM.II.K.S11)

25. **Measure the output voltage of a DC generator.** (AM.II.K.S11)

26. **Check the resistance of an electrical system component provided by the examiner.** (AM.II.K.S12)

27. **Check generator brushes for wear and brush springs for proper tension.** (AM.II.K.S13)

28. **Inspect and check anti-collision, position, and/or landing lights for proper operation.** (AM.II.K.S14)

29. **Identify and inspect the components in an electrical system.** (AM.II.K.S15)

30. **Troubleshoot a DC electrical system supplied by an alternating current (AC) electrical system.** (AM.II.K.S16)

31. **Identify the components in an electrical schematic where AC is rectified to a DC voltage.** (AM.II.K.S17)

32. **Demonstrate to the examiner how to perform a continuity test on a conductor.** (AM.II.K.S18)

33. **Demonstrate to the examiner how to perform a test on a conductor for a short to ground.** (AM.II.K.S19)

34. **Demonstrate to the examiner how to perform a test on a conductor for a short to other conductors.** (AM.II.K.S20)

L. Ice and Rain Control Systems

References: AC 43.13-1; FAA-H-8083-31

Knowledge

1. **Why is it important that ice not be allowed to build up on airplane wings in flight?** (AM.II.L.K1)

   Ice distorts the shape of the airfoil and destroys the aerodynamic lift. The weight of the ice loads the aircraft down.

2. **What is the function of an ice detector when ice is detected?** (AM.II.L.K2)

   To illuminate an annunciator light in the flight deck and, in some aircraft, to automatically turn on anti-icing systems.

3. **How is ice kept from forming on the pitot tube of an airplane?** (AM.II.L.K3)

   Pitot tubes are heated by electric current flowing through heater elements built into them.

4. **How is ice prevented from forming on the windshield of modern jet transport airplanes?** (AM.II.L.K3)

   The windshield has a heater element embedded in it. Electric current heats the windshield and keeps ice from forming on it.

5. **Are pneumatic deicer boots operated before ice forms or after it has formed?** (AM.II.L.K4)

   Pneumatic deicer boots are not operated until ice has formed over them. When the boot inflates, it breaks the ice, and the air flowing over the airfoil blows it away.

6. **Where does the air come from to operate the pneumatic deicer boots on a reciprocating-engine powered airplane?** (AM.II.L.K4)

   From the discharge side of the air pump used to operate the gyro instruments.

7. **What is meant by a wet vacuum pump used to power pneumatic deicer boots?** (AM.II.L.K4)

   It is a vacuum pump that uses engine oil to lubricate its steel vanes. A dry vacuum pump uses carbon vanes, and it does not require any oil for lubrication.

8. **What is the purpose of the oil separator in a deicer system?** (AM.II.L.K4)

   Oil separators are used with wet vacuum pumps to remove the lubricating oil from the discharge air before this air is used in the deicer boots.

9. **How are rubber deicer boots cleaned?** (AM.II.L.K4)

   By washing them with mild soap and water.

10. **How are rubber deicer boots attached to the leading edges of aircraft wings and tail surfaces?** (AM.II.L.K4)

    They are bonded to the surface with an adhesive. Boots on the older aircraft were attached with machine screws and Rivnuts.

11. **What is the sequence of inflation of a deicing boot system?** (AM.II.L.K4)

    There are three tubes in a deicer boot, upper, center, and lower. The center tube inflates first with the upper and lower tubes a specified number of seconds later.

12. **When should rain repellent be used on an airplane windshield?** (AM.II.L.K5)

    Only when the windshield is wet with rain.

13. **Where can you find the instructions for adjusting the tension for the windshield wiper blades?** (AM.II.L.K5)

    In the aircraft maintenance manual.

14. **How can you tell, on a preflight inspection, that the pitot heater is operating?** (AM.II.L.K6)

    Turn it on for about a minute and then on the walk-around inspection, feel it to see that it is warm.

15. **Where can you find the correct part number for the electrically heated pitot head used on an aircraft?** (AM.II.L.K6)

    In the illustrated parts manual for the aircraft.

16. **If the pitot head includes ports for the static air system, what must be done after the head is replaced?** (AM.II.L.K6)
    A static system leak test must be performed and the results recorded in the aircraft maintenance records.

17. **How can you test if the pitot heat is working?** (AM.II.L.K6)
    Briefly turn the pitot heat system on and observe an increased current indication on the ammeter.

18. **What might be likely causes for a propeller electrothermal deicer system failure?** (AM.II.L.K6)
    A tripped circuit breaker or broken wire to the electrical heating elements.

19. **How can a pilot determine if an electrothermal propeller deicer is working?** (AM.II.L.K6)
    The pilot should observe an increased ammeter reading when the system is energized.

20. **What are two ways rain can be kept from obstructing the pilot's vision through the windshield of an airplane?** (AM.II.L.K7)
    The rain can be blown away by a high velocity blast of compressor bleed air, or it can be wiped away with electrically or hydraulically operated windshield wipers.

Risk Management

1. **What risks are associated with testing and maintenance of ice control systems?** (AM.II.L.R1)
   Running windshield wipers on a dry windshield can damage the windshield. Over-pressurizing deice boots with a ground test system can damage the boots. Follow the maintenance manual procedures whenever testing or servicing ice control systems.

2. **What are the dangers of improper handling and storage of deicing fluids?** (AM.II.L.R2)
   Deice fluids can be hazardous to health and the environment. Always follow the manufacturer's recommendations for handling and storage.

3. **What are the risks associated with using incorrect cleaning materials on heated windshields?** (AM.II.L.R3)
   Using the incorrect cleaning materials can damage windshields and reduce visibility through the windshield. Follow the maintenance manual procedures for cleaning heated windshields.

Skills

1. **Actuate a pitot heater and check it for proper operation.** (AM.II.L.S1)

2. **Locate the part number for a pitot head and explain to the examiner the way the head would be replaced.** (AM.II.L.S1)

3. **Perform an operational check of the replaced pitot head.** (AM.II.L.S1)

4. **Make an entry of the replaced pitot head in the aircraft maintenance records.** (AM.II.L.S1)

5. **Demonstrate to the examiner the correct way to patch a pneumatic deicer boot following the instructions of the boot manufacturer.** (AM.II.L.S2)

6. **Perform an operational check of the deicer system, explaining to the examiner the purpose of the sequencing of the inflation of the boots.** (AM.II.L.S2)

7. **Inspect a pneumatic deicer boot.** (AM.II.L.S2)

8. **Clean a pneumatic deicer boot.** (AM.II.L.S3)

9. **Locate the procedures for troubleshooting an electrically heated pitot system.** (AM.II.L.S4)

10. **Troubleshoot an electrically heated pitot system provided by the examiner.** (AM.II.L.S4)

11. **Check an electrically heated water drain system.** (AM.II.L.S4)

12. **Inspect the thermal anti-ice systems.** (AM.II.L.S5)

13. **Inpect and operational check an electrically heated windshield.** (AM.II.L.S6)

14. **Examine the blades of a windshield wiper system, and check them for the correct tension and for the correct parking position.** (AM.II.L.S7)

15. **Check an electrically or hydraulically operated windshield wiper system.** (AM.II.L.S7)

16. **Inspect an electrically operated windshield wiper system.** (AM.II.L.S7)

17. **Replace blades on a windshield wiper system.** (AM.II.L.S8)

18. **Locate and explain the procedures for replacing blades on a windshield wiper system.** (AM.II.L.S8)

19. **Check pneumatic rain removal system.** (AM.II.L.S9)

20. **Locate and explain the procedures for inspecting a pneumatic rain removal system.** (AM.II.L.S9)

M. Airframe Fire Protection Systems

References: AC 43.13-1; FAA-H-8083-31

Knowledge

1. **What is a Class B fire?** (AM.II.M.K1)
   Class B fires are those caused by combustible fluids.

2. **What are the three elements required for a fire?** (AM.II.M.K1)
   Fuel, oxygen, and high enough temperature.

3. **What is a fire zone on an aircraft?** (AM.II.M.K1)
   A fire zone is an area or region of an aircraft designed by the manufacturer to require fire detection and/or fire extinguishing equipment and a high degree of inherent fire resistance.

4. **What type of fire detection system is usually used in engine compartments, APU installations, and wheel wells?** (AM.II.M.K2)
   A continuous-loop fire detection system.

5. **Does a thermocouple fire detection system warn the pilot of a general overheat condition?** (AM.II.M.K2)
   No, it operates on the rate of temperature rise, and it detects only a fire.

6. **Does a thermal switch fire detection warn the pilot of a general overheat condition?** (AM.II.M.K2)
   No, it actuates only when there is a fire.

7. **Does a pneumatic fire detection system warn the pilot of a general overheat condition?** (AM.II.M.K2)
   Yes, it actuates when there is a fire or a general overheat condition.

8. **How is a thermoswitch fire detector circuit checked?** (AM.II.M.K3)
   Close the test switch. If the system is continuous and not shorted, the fire-warning light and bell will actuate. Failure of the warning light to illuminate shows the system is faulty.

9. **How is a continuous-loop fire detector circuit checked?** (AM.II.M.K3)
   Close the test switch. If the system is continuous and not shorted, the fire-warning light and bell will actuate. Failure of the warning light to illuminate shows the system is faulty.

10. **How is the pneumatic fire detection system tested?** (AM.II.M.K3)
    When the test switch is closed, low-voltage AC flows through the stainless steel tube in which the gas absorbing element is housed. This current heats the element which releases gas and actuates the fire warning light and bell.

11. **What type of detector is commonly used in baggage compartments and holds?** (AM.II.M.K4)
    Smoke detectors.

12. **What type of detector is usually used in the flight deck and cabin areas?** (AM.II.M.K4)
CO detectors.

13. **What type of fire extinguishing agent is best for both cabin fires and engine fires?** (AM.II.M.K5)
Halon 1301.

14. **What types of fire extinguishing agents should not be used in aircraft cabins or flight decks? (AM.II.M.K5)**
CO_2, dry chemicals, or specialized dry powders.

15. **What is a major disadvantage of "CB" fire extinguishing agent for extinguishing aircraft fires?** (AM.II.M.K5)
It is corrosive to aluminum and magnesium.

16. **Why is carbon tetrachloride not recommended as a fire extinguishing agent?** (AM.II.M.K5)
Carbon tetrachloride produces phosgene, a deadly gas, when it is exposed to flames.

17. **What is used as a fire extinguishing agent in most of the high-rate discharge systems installed in aircraft?** (AM.II.M.K5)
One of the halogenated hydrocarbons, such as Halon 1301, pressurized with nitrogen.

18. **What releases the fire extinguishing agent in a high-rate discharge bottle?** (AM.II.M.K6)
An electrically ignited powder charge blows a knife through a seal in the HRD bottle.

19. **What happens when the Fire-Pull T-handle is pulled in a jet transport aircraft?** (AM.II.M.K6)
The bottle discharge switch is uncovered and armed, the generator field relay is tripped, fuel is shut off to the engine, and hydraulic fluid is shut off to the pump. The engine bleed air is shut off and the hydraulic pump low-pressure lights are deactivated.

20. **What must be done to release the fire extinguishing agent after the Fire-Pull T-handle has been pulled?** (AM.II.M.K6)
The Bottle Discharge switch must be closed.

21. **What type of fire extinguisher is recommended for extinguishing a brake fire?** (AM.II.M.K6)
A dry-powder type extinguisher.

22. **How can you determine the state of charge of a CO_2 fire extinguisher?** (AM.II.M.K7)
By its weight.

23. **How can you determine the state of charge of a freon fire extinguisher container?** (AM.II.M.K7)
By the pressure shown on the built-in gauge.

24. **How does the ambient temperature affect the pressure shown on the pressure gauge on a freon fire extinguisher?** (AM.II.M.K7)

    The higher the temperature, the higher the pressure.

25. **How can you determine whether or not a built-in fire extinguishing system has been discharged?** (AM.II.M.K7)

    By checking the blowout plugs on the outside of the aircraft near the extinguisher agent bottles.

26. **What is indicated if the red disc in a built-in fire extinguishing system is blown out?** (AM.II.M.K7)

    The agent bottle has been discharged because of an overheat condition.

27. **What is indicated if the yellow disc in a built-in fire extinguishing system is blown out?** (AM.II.M.K7)

    The agent bottle has been discharged by the flight crew actuating the Bottle Discharge switch.

Risk Management

1. **What precaution must you observe when checking the electrical squib of an HRD fire extinguisher bottle for electrical continuity?** (AM.II.M.R1)

   It takes only a small amount of current to ignite the powder charge, and the method of testing must not send this amount of current through it.

2. **Why is it important to use proper PPE when working on or testing fire extinguishing systems?** (AM.II.M.R2)

   Some fire extinguishing systems operate under very high pressures, using chemicals that can be hazardous to one's health. Proper PPE should always be used to protect the individuals working on the systems.

3. **What are the risks of using improper fire extinguishing agents?** (AM.II.M.R3)

   The use of improper fire extinguishing agents can be ineffective for the fire at hand. In some cases, use of an incorrect agent in an enclosed area may make it difficult for individuals to breathe.

Skills

1. **Inspect, check, troubleshoot, and/or repair a fire detection system.** (AM.II.M.S1)

2. **Given a schematic diagram of a fire detection system, describe to the examiner faults that could cause a false alarm, and faults that could prevent the system indicating the presence of a fire.** (AM.II.M.S1)

3. **Locate the troubleshooting procedures for a high rate of discharge fire extinguisher system.** (AM.II.M.S1)

4. **Demonstrate to the examiner the correct way to check the pressure on a high-rate discharge fire extinguisher container. Explain the reason for the allowable pressure range.** (AM.II.M.S2)

5. **Determine the proper container pressure for an installed fire extinguisher system.** (AM.II.M.S2)

6. **Demonstrate to the examiner the correct way to remove and replace an HRD fire extinguishing agent container.** (AM.II.M.S3)

7. **Demonstrate to the examiner where to find the maintenance procedures for the assigned fire detection and extinguishing system and components.** (AM.II.M.S3)

8. **Inspect a smoke and toxic gas detection system assigned by the examiner.** (AM.II.M.S4)

9. **Locate the inspection procedures for carbon monoxide detectors and perform the required inpsections.** (AM.II.M.S5)

10. **Locate and explain the procedures for checking a smoke detection system.** (AM.II.M.S6)

11. **Locate and explain the procedures for inspecting an overheat detection system.** (AM.II.M.S7)

12. **Demonstrate to the examiner the correct way to determine the state of charge of a CO_2 fire extinguisher bottle.** (AM.II.M.S8)

13. **Inspect the fire protection system CO_2 cylinders.** (AM.II.M.S8)

14. **Inspect a conventional CO_2 fire protection system.** (AM.II.M.S8)

15. **Check a conventional CO_2 fire protection system.** (AM.II.M.S8)

16. **Inspect a fire-extinguisher bottle or cylinder for hydrostatic test date.** (AM.II.M.S8)

17. **Locate the procedures for inspecting a thermal switch fire detection system.** (AM.II.M.S9)

18. **Inspect a thermocouple fire warning system.** (AM.II.M.S9)

19. **Inspect an HRD fire extinguishing system for security of its components and for the state of charge of its bottles.** (AM.II.M.S9)

20. **Explain to the examiner the way the electrical circuit for an HRD fire extinguishing system should be checked.** (AM.II.M.S9)

21. **Locate and identify the blow-out plugs that indicate the status of the fire extinguisher system in an aircraft.** (AM.II.M.S9)

22. **Perform an operational check of a fire detection/protection system assigned by the examiner.** (AM.II.M.S10)

23. **Check a Freon bottle discharge circuit.** (AM.II.M.S11)

24. **Check a fire protection system Freon bottle charge pressure.** (AM.II.M.S11)

25. **Inspect a Freon bottle discharge cartridge.** (AM.II.M.S11)

26. **Check a continuous loop fire detection system.** (AM.II.M.S12)

27. **Inspect a continuous loop fire detection system.** (AM.II.M.S12)

N. Rotorcraft Fundamentals

References: AC 43.13-1; FAA-H-8083-31

Knowledge

1. **What causes dissymmetry of lift produced by the rotor of a helicopter?** (AM.II.N.K1)
   The forward speed of the helicopter produces dissymmetry of lift. The rotor blade that is traveling forward as the helicopter is flying produces more lift than the blade that is traveling rearward.

2. **Why do single-rotor helicopters use an auxiliary rotor on their tail?** (AM.II.N.K2)
   The thrust from the tail rotor counteracts the torque produced by the main rotor.

3. **Why is it important that the blades of a helicopter rotor system be in track?** (AM.II.N.K2)
   If the blades are not in track, vertical vibration can develop.

4. **What is the function of the collective pitch control of a helicopter?** (AM.II.N.K2)
   It changes the pitch of all the blades at the same time. The collective pitch controls the vertical flight of the helicopter.

5. **What is the function of the cyclic pitch control?** (AM.II.N.K2)
   It changes the pitch of the rotor blades at a particular point in their rotation to tilt the plane of the rotor. The cyclic pitch controls the lateral movement of the helicopter.

6. **What is the function of the tail rotor on a single main rotor helicopter?** (AM.II.N.K2)
   The thrust from the tail rotor counteracts the torque of the main rotor to control the yaw of the helicopter. The control pedals change the pitch of the tail rotor blades to vary the thrust.

7. **What is the purpose of the stabilizer system in a helicopter?** (AM.II.N.K2)
   A helicopter is statically stable but dynamically unstable. The stabilizer system restores the helicopter to level flight when outside forces cause it to pitch or roll.

8. **What is the purpose of a transmission on a helicopter?** (AM.II.N.K3)
   The transmission system transfers power from the engine to the main rotor, tail rotor, and other accessories during normal flight conditions.

9. **What are the basic flight control rigging procedures for a helicopter?** (AM.II.N.K4)
   1. Place the controls in a set position and clamp or pin in place,
   2. place the control surfaces in the specified position using rigging jigs or a precision protractor, and
   3. set the maximum travel stops for the controls.

10. **What is meant by a fully articulated rotor system?** (AM.II.N.K5)
    It is a rotor system in which the individual blades are free to flap, drag, and feather.

11. **What is meant by a semirigid rotor system?** (AM.II.N.K5)
    A two-blade rotor system in which the blades can feather, but cannot flap or drag. The rotor hub is mounted on the mast with a teetering hinge that allows the entire rotor to rock as a unit.

12. **What is meant by a rigid rotor system?** (AM.II.N.K5)
    A rotor system that has freedom of motion about its feather axis only. The flexibility of the blades is sufficient to provide the needed flapping and dragging.

13. **What is the purpose of a helicopter skid shoe?** (AM.II.N.K6)
    To protect the helicopter landing skid tubes from wear.

14. **What is the shape of a cross section of a rotor blade?** (AM.II.N.K7)
    The shape of a rotor blade is an airfoil, similar to that of an aircraft wing.

15. **What type of construction is used in rotor blades?** (AM.II.N.K7)
    Rotor blades can be constructed of aluminum extrusions and sheet metal, composite construction, or a combination of metallic and composite components.

16. **What is the basic cause of low-frequency lateral vibration?** (AM.II.N.K8)
    The main rotor blades being out of static balance.

17. **What is the basic cause of low-frequency vertical vibration?** (AM.II.N.K8)
    One of the main rotor blades producing more lift than the other.

18. **What is usually the cause of high-frequency vibration?** (AM.II.N.K8)
    The engine, cooling fan, or tail rotor.

19. **What is the most effective way to check a helicopter rotor for dynamic balance?** (AM.II.N.K8)
    Use a special computerized analyzer/balancer.

20. **What is the purpose of transmission mount dampening?** (AM.II.N.K9)
    Extreme low-frequency vibration is associated with pylon rock (two or three cycles per second) and is inherent with the rotor, mast, and transmission system. Transmission mount dampening is incorporated to absorb the rocking.

Risk Management

1. **What are the risks associated with working around helicopter blades during ground operations?** (AM.II.N.R1)
   Main rotor blades may dip, and the tail rotor is exposed on some helicopters, both of which may cause physical harm to personnel. Training of personnel and carefully followed safety procedures are critical for safe operations.

2. **What are the risks associated with the ground handling of helicopters?** (AM.II.N.R2)
   Improper ground handling procedures can cause damage to the helicopter, other aircraft, or personnel. Aviation mechanics are to follow ground handling procedures found in the maintenance manual.

3. **What are the risks associated with improper ground operations and functional tests?** (AM.II.N.R3)
   The risks are total loss of the aircraft and injury to personnel. Always follow the instructions and procedures found in the maintenance manual.

4. **Why is it important to follow maintenance manual procedures during maintenance and inspection of rotorcraft systems and components?** (AM.II.N.R4)
   The proper maintenance of helicopters is vital to their safety and airworthiness. Many systems have close tolerances and must be rigged correctly for safe operations.

Skills

1. **Demonstrate to the examiner the correct way to check a helicopter rotor for track.** (AM.II.N.S1)

2. **Locate and explain to the examiner the main components of a helicopter rotor system.** (AM.II.N.S1)

3. **Explain to the examiner the operation of the swashplate of a helicopter rotor system.** (AM.II.N.S1)

4. **Demonstrate to the examiner the way to check the control rods for proper installation and safetying.** (AM.II.N.S2)

5. **Demonstrate to the examiner where to find the helicopter rotor blade track and balance procedures.** (AM.II.N.S2)

6. **Locate the causes of vertical vibration in a two-rotor helicopter rotor system.** (AM.II.N.S3)

7. **Locate and explain procedures needed to rig the specified helicopter controls.** (AM.II.N.S3)

8. **Locate and explain procedures to track and balance a rotor system.** (AM.II.N.S4)

O. Water and Waste Systems

References: AC 25.1455-1, AC 120-39; FAA-H-8083-31

Knowledge

1. **Is potable water on aircraft suitable for drinking?** (AM.II.O.K1)
   Yes, potable water is safe for drinking.

2. **What are the main components of a potable water system?** (AM.II.O.K1)
   Water is stored in tanks that hold 30 to 50 gallons. The system has servicing ports for filling and draining, monitoring and pressurization systems, water heaters, and distribution lines.

3. **What is potable water used for on an aircraft?** (AM.II.O.K1)
   To supply fresh water for galleys and lavatories.

4. **What are the two main types of lavatory waste systems?** (AM.II.O.K2)
   The closed waste system (or recirculating system) and the vacuum waste system.

5. **What are the inspection and servicing requirements for potable water and waste systems?** (AM.II.O.K3)
   a. Perform an operational check on all components.
   b. Verify adequate water pressure and draining at all sinks.
   c. Check for hot water (if installed).
   d. Flush each lavatory, observing operation and correct cycling and timing of the components.
   e. Inspect all components and lines for leaks.
   f. Inspect for corrosion in the surrounding structure, especially below waste tanks.
   g. Inspect all control panels and indicators for proper operation.

Risk Management

1. **What precautions should be taken when servicing lavatory waste systems?** (AM.II.O.R1)
   Waste system pre-charge fluids contain chemicals that are hazardous. Use the proper PPE and use the specified equipment and procedures when servicing the system.

Skills

1. **Locate and explain the procedures for servicing a lavatory waste system.** (AM.II.O.S1)

2. **Locate and explain the procedures for servicing a potable water system.** (AM.II.O.S2)

The Powerplant Oral and Practical Tests

There are 13 subject areas that are tested on the Powerplant Oral and Practical Exams.

For each subject area, this guide provides typical oral questions and succinct answers for the knowledge and risk management elements and presents the skills that applicants must understand and demonstrate.

III. Powerplant

A. Reciprocating Engines
B. Turbine Engines
C. Engine Inspection
D. Engine Instrument Systems
E. Engine Fire Protection Systems
F. Engine Electrical Systems
G. Engine Lubrication Systems
H. Ignition and Starting Systems
I. Engine Fuel and Fuel Metering Systems
J. Reciprocating Engine Induction and Cooling Systems
K. Turbine Engine Air Systems
L. Engine Exhaust and Reverser Systems
M. Propellers

A. Reciprocating Engines

References: 14 CFR Part 43; AC 43.13-1; FAA-H-8083-32

Knowledge

1. **What are the basic types or configurations of reciprocating engines?** (AM.III.A.K1)
   Inline, opposed, v-type, and radial.

2. **In what position should the crankshaft be when a magneto is being timed to the engine?** (AM.III.A.K2)
   The piston in cylinder number 1 should be in the correct position for ignition to occur. This is normally about 30° before top center on the compression stroke.

3. **What is checked on engine runup to determine the operational condition of the engine?** (AM.III.A.K2)
   Idling RPM and manifold pressure, engine acceleration, maximum static RPM and manifold pressure, magneto drop, ignition switch safety check, propeller pitch change, propeller feathering, oil pressure, fuel pressure, and fuel flow.

4. **What is a hydraulic lock in an aircraft engine?** (AM.III.A.K3)
   Oil has drained past the piston rings and filled the combustion chamber of cylinders that are below the center line of the engine. If the crankshaft is turned over with this oil in the cylinder it will lock, and there is a probability that the piston, cylinder, or connecting rod will be damaged.

5. **How is a hydraulic lock removed from an aircraft engine?** (AM.III.A.K3)
   Remove a spark plug from the locked cylinder and drain all the oil out. Clean the spark plug and replace it.

6. **What are the strokes of a 4-stroke reciprocating engine?** (AM.III.A.K3)
   Intake, compression, power, and exhaust.

7. **What are two main causes of a cylinder having low compression?** (AM.III.A.K4)
   Piston rings not seating or broken, and valves not seating or burned.

8. **What is a low-tension ignition system?** (AM.III.A.K4)
   It is an ignition system using a magneto with a brush-type distributor and a coil that has no secondary wiring. Low voltage is directed from the magneto to a transformer mounted on the cylinder head. Here the low voltage is boosted to a high voltage and is directed through a very short high-tension lead to the spark plug. There is one transformer for each spark plug.

9. **What are some advantages of a horizontally opposed engine?** (AM.III.A.K4)
   Low power-to-weight ratio, ideal for installation on aircraft wings, and low vibration characteristics.

10. **How can you determine the firing order of an 18-cylinder radial engine?** (AM.III.A.K5)

    Begin with cylinder 1, and add 11 or subtract 7, whichever will keep the numbers between 1 and 18. 1-12-5-16-9-2-13-6-17-10-3-14-7-18-11-4-15-8.

11. **What is a floating cam ring on a large radial engine?** (AM.III.A.K5)

    It is a cam ring that rides on a shelf-type bearing with a large amount of clearance between the bearing and the ring. When adjusting the valves on an engine with a floating cam ring, the pressure of the valve springs on the opposite side of the engine must be removed so the cam ring will be tight against the bearing for the valves being adjusted.

12. **Why is valve clearance adjusted on radial engines, but not on most horizontally opposed engines?** (AM.III.A.K5)

    Most horizontally opposed engines have hydraulic valve lifters that keep all of the clearance out of the valve train. Radial engines have solid lifters.

13. **Why are both the hot and cold valve clearances given for most radial engines?** (AM.III.A.K5)

    The hot clearance is given for valve timing purposes. The timing is adjusted with the valves in cylinder number one, set with the hot clearance. When the timing is set, all of the valves are adjusted to their cold clearance.

14. **Why is it important to preserve engines that will not be operated for an extended period of time?** (AM.III.A.K6)

    Moisture will accumulate in the engine, initiating the corrosion process which can ruin the engine.

15. **Preservation for temporary storage is valid for engines that will be out of service for how many days?** (AM.III.A.K6)

    30 to 90 days.

16. **What should be done to the cylinders to protect them from rust and corrosion when preparing the engine for long-time storage?** (AM.III.A.K6)

    Spray a preservative oil inside the cylinders and replace the spark plugs with desiccant plugs.

17. **What is the formula for calculating indicated horsepower on a 4-stroke engine?** (AM.III.A.K7)

    $$\text{Indicated horsepower} = \frac{(PLANK)}{33{,}000}$$

    Where:
    P = Indicated mean effective pressure, in psi
    L = Length of the stroke, in feet or in fractions of a foot
    A = Area of the piston head or cross-sectional area of the cylinder, in square inches
    N = Number of power strokes per minute: RPM ÷ 2
    K = Number of cylinders

18. **What is specific fuel consumption (SFC)?** (AM.III.A.K7)

SFC is a measure of the economical use of fuel, which is the fuel weight in pounds per hour per horsepower:

$$SFC = \frac{\text{(pounds fuel/hour)}}{\text{horsepower}}$$

19. **What must be inspected to determine the airworthiness of an engine installation?** (AM.III.A.K8)

The propeller, lubrication system, ignition system, fuel metering system, cooling system, and the exhaust system.

20. **Where can you find the correct grade of fuel and amount of engine lubricating oil for an aircraft engine?** (AM.III.A.K8)

In the Type Certificate Data Sheets for the aircraft.

21. **What are the basic steps in troubleshooting an engine installation?** (AM.III.A.K8)

a. Know how the system should operate.
b. Observe how the system is operating.
c. Divide the system into smaller parts to find the trouble.
d. Look for the obvious problems first.

22. **Where do you find instructions for removing an engine from an aircraft?** (AM.III.A.K8)

In the aircraft maintenance manual.

23. **What must normally be done to a nose-wheel airplane before removing an engine?** (AM.III.A.K8)

The tail of the airplane must be supported.

24. **What should be used to cover the ends of fuel and oil lines that have been disconnected from the engine?** (AM.III.A.K8)

The correct plugs or caps. Never cover the end of a line with tape.

25. **What is meant by a QEC engine assembly?** (AM.III.A.K8)

An engine that has been prepared for a quick engine change. The engine is on its mount and all of the accessories are installed. A minimum of time is required to change engines. All that is needed is for the controls, wiring, and fluid lines to be connected to the firewall.

26. **What is meant by rigging the engine controls so they have some "cushion"?** (AM.III.A.K8)

The control on the engine component must contact its stop before the control handle in the flight deck reaches its stop. This causes the control handle to spring back a slight amount when it is moved to the extent of its travel.

27. **What are two ways of pre-oiling an engine?** (AM.III.A.K8)

a. Before the spark plugs are installed and after the oil tank or sump is filled, turn the engine over with the starter until oil runs out of the fitting to which the oil pressure gauge connects.
b. Use a pre-oiler tank. Air pressure forces oil through all of the passages until some of it flows from the oil pressure gauge fitting.

28. **What checks should be made to determine the proper operation of a reciprocating engine installation?** (AM.III.A.K9)

    Idle RPM and mixture, static RPM, propeller pitch change operation, magneto check, ignition switch safety check, and carburetor heat check.

29. **What is the purpose of a cold cylinder check?** (AM.III.A.K9)

    A cold cylinder check determines which cylinder is not firing. Run the engine for a few minutes at the speed, and on the magneto, at which it runs roughest. Shut it down and feel the exhaust stack near the cylinder head. The cylinder with the stack that is not as hot as the others is the cylinder that has not been firing.

30. **How is fuel ignited in a diesel cycle engine?** (AM.III.A.K10)

    The diesel cycle depends on high compression pressures to provide for the ignition of the air-fuel charge in the cylinder.

Risk Management

1. **What are the risks of moving a propeller by hand during aircraft maintenance?** (AM.III.A.R1)

   If the ignition switch is on, or a magneto ground wire is faulty, the engine can start if the impulse coupling is fired when the propeller is turned. Maintenance personnel should always verify the magneto switch is off and stay clear of the propeller arc any time the propeller is moved.

2. **What precautions should be taken prior to and during ground operations of an engine?** (AM.III.A.R2)

   Improper procedures during ground runs can cause damage to the aircraft and injury to personnel. The maintenance manual and operational checklists should always be used when operating an engine and aircraft on the ground.

3. **What are the risks associated with incorrect procedures in the event of a reciprocating engine fire?** (AM.III.A.R3)

   Improper procedures can lead to further damage to the engine and aircraft. If the engine backfires and causes a fire in the induction system, the starter motor should continue to run while mixture control is pulled to idle cutoff to prevent more fuel from entering the engine. If the fire is outside of the engine, pull the mixture control to idle cutoff and close the fuel firewall shutoff valve to stop the flow of fuel into the engine compartment. After aircraft maintenance, it is good practice to have a spotter standing by with a fire extinguisher.

4. **What risks are associated with using procedures other than those developed by the manufacturer?** (AM.III.A.R4)

   The risk of using other procedures is that important steps and procedures can be missed, causing damage to the engine, aircraft, or personnel. The manufacturer's procedures were carefully developed to maintain the type design configuration and safety of the aircraft.

Skills

1. **Demonstrate the correct way to replace the packing seals around a push rod housing.** (AM.III.A.S1)

2. **Explain to the examiner the way a cylinder should be inspected and repaired if the differential compression check identifies a leak past the valves.** (AM.III.A.S1)

3. **Perform a cold cylinder check.** (AM.III.A.S1)

4. **Where do you find the procedures to follow when correcting a defective condition in an aircraft engine?** (AM.III.A.S2)

   In the engine maintenance manual.

5. **Perform an engine runup to determine that the engine is operating according to the engine manufacturer's specifications. Explain to the examiner what should be checked on this runup.** (AM.III.A.S2)

6. **Troubleshoot faults found in a reciprocating engine.** (AM.III.A.S2)

7. **Install the piston and/or knuckle pin(s).** (AM.III.A.S3)

8. **Identify the parts of a cylinder provided by the examiner.** (AM.III.A.S4)

9. **Identify the parts of a crankshaft and perform a dimensional inspection.** (AM.III.A.S5)

10. **Identify and inspect various types of bearings.** (AM.III.A.S6)

11. **Check all engine controls for proper travel and free motion. Explain to the examiner the reason for rigging the controls so they have some cushion.** (AM.III.A.S7)

12. **Locate top dead-center on an engine provided by the examiner.** (AM.III.A.S8)

13. **Remove a cylinder. Inspect the cylinder and identify the parts and describe your findings to the examiner. Properly reinstall the cylinder.** (AM.III.A.S9)

B. Turbine Engines

References: 14 CFR Part 43; AC 43.13-1; FAA-H-8083-32

Knowledge

1. **How are the four strokes of a reciprocating engine similar to the sections of a turbine engine?** (AM.III.B.K1)
   Intake = air inlet section, compression = compressor section, ignition and power = combustion and turbine sections, exhaust = exhaust section.

2. **What type of turbojet engine is commonly used in commercial aviation due to lower noise and better fuel consumption?** (AM.III.B.K1)
   High-bypass turbofan engines.

3. **What four types of gas turbine engines are used in aircraft?** (AM.III.B.K2)
   Turbofan, turboprop, turboshaft, and turbojet.

4. **What are the two types of turbine compressors?** (AM.III.B.K3)
   Centrifugal flow and axial flow.

5. **What is meant by trimming a turbine engine?** (AM.III.B.K4)
   Adjusting the fuel control so the engine develops the correct idle and trim speed RPM.

6. **What type of equipment should be used to determine that a turbine engine is performing up to the standards specified by the engine manufacturer?** (AM.III.B.K4)
   A JetCal analyzer/trimmer has all of the instrumentation and cables to determine the EGT, EPR, and other parameters specified by the engine manufacturer.

7. **What is meant by a hot-section inspection?** (AM.III.B.K5)
   An inspection of the hot section of a turbine engine. The hot section includes the combustors, the turbine, and the exhaust system.

8. **What is meant by on-condition maintenance of a turbine engine?** (AM.III.B.K5)
   The monitoring of the engine performance at regular intervals and determining when maintenance is required based on certain operating parameters specified by the engine manufacturer.

9. **What checks of a turbine engine are necessary to verify preflight condition?** (AM.III.B.K5)
   a. Inspect the cowling and all air inlet areas and attachment of the engine to the airframe.
   b. Inspect the inlet guide vanes and the first stage of the compressor or fan blades.
   c. Check for unusual noise when the compressor is rotated.
   d. Inspect the rear turbine and exhaust system.
   e. Check the quantity of the lubricating oil for the engine and the constant-speed drive unit.
   f. Check the ignition system for operation by listening for the sparks.

10. **What engine instruments are monitored during the start of a turbine engine?** (AM.III.B.K6)

    Tachometer, oil pressure, and internal gas temperatures.

11. **What checks are performed after the installation of a turbine engine?** (AM.III.B.K6)

    After starting and idle speed is obtained, the engine should stabilize, and engine instrument values are monitored. Takeoff thrust is checked by adjusting the throttle to obtain a single, predicted reading on the engine pressure ratio indicator. The manufacturer's procedures and limitations must be followed during all engine operations. After engine shut down, inspect the engine for fuel and oil leaks and general condition.

12. **What engine fault could cause a turbine engine to have high exhaust gas temperature, low RPM, and high fuel flow at all engine pressure ratio settings?** (AM.III.B.K7)

    Possible turbine damage and/or loss of turbine efficiency due to excessive wear.

13. **What would be the indications of an engine bleed-air valve malfunction?** (AM.III.B.K7)

    Engine has higher-than-normal exhaust gas temperature during takeoff, climb, and cruise. RPM and fuel flow are higher than normal.

14. **What is the primary purpose of bleeding air off the compressor section?** (AM.III.B.K8)

    To unload the compressor at low speeds and to control the flow of air to the combustion burners.

15. **What is the secondary purpose or use of bleed air?** (AM.III.B.K8)

    Bleed air is used in a variety of aircraft systems including cabin pressurization, heating and cooling, deicing and anti-icing systems, pneumatic starting of engines and auxiliary drive units, and vortex dissipaters.

16. **How is a turbine engine preserved for long periods of inactivity?** (AM.III.B.K9)

    The methods will vary depending on the manufacturer and the length of inactivity. Common practices include draining the lubrication system and flushing with preservative oil. The engine fuel system may also be drained and filled with preservative oil, including the fuel control. Before the engine can be returned to service, the preservative oil must be completely flushed from the fuel system by motoring the engine and bleeding the fuel system. Always follow the manufacturer's instructions when performing any preservation or de-preservation of gas turbine engines.

17. **Where are most APUs located in modern jet transport aircraft?** (AM.III.B.K10)

    In the tail cone of the fuselage.

18. **What is the function of an APU?** (AM.III.B.K10)

    An APU provides electric power and compressed air when the main engines are not operating.

19. **What are two sources of compressed air from an APU?** (AM.III.B.K10)

    From bleed air from the APU turbine compressor, or from a load compressor driven by a free turbine in the engine.

20. **Where in maintenance information based on the ATA-100 system would you find instructions for inspecting and servicing the engine of an airborne APU?** (AM.III.B.K10)

    In section 49-20.

21. **How is most of the troubleshooting done for a modern APU?** (AM.III.B.K10)

    By the fault codes generated by the FADEC.

22. **What ensures that an APU will not be shut down while it is too hot?** (AM.III.B.K10)

    The APU fuel control incorporates an automatic time-delay feature that closes the bleed air valve to remove most of the load and reduce the APU temperature before it is shut down.

23. **What prevents the APU exceeding its safe operating limits when the bleed air valve is wide open?** (AM.III.B.K10)

    The FADEC monitors the load and regulates the fuel going to the APU to prevent it from exceeding its safe limits.

24. **What document would you use to find the safety procedures to follow when replacing an igniter plug in an APU engine?** (AM.III.B.K10)

    The maintenance manual for the APU.

25. **What adjustments and rigging are accomplished on a turbine engine?** (AM.III.B.K11)

    The primary rigging on a turbine engine involves the adjustment of power control levers and cables, followed by careful adjustment of the fuel control unit to trim the engine for the correct idle speeds and to obtain maximum thrust.

Risk Management

1. **What are the risks associated with operating a turbine engine?** (AM.III.B.R1)

    Improper engine operating procedures can be very expensive, and damage can happen quickly, with repair costs of hundreds of thousands of dollars or more. Turbine engines can also create enormous amounts of thrust, and the jet blast from engines can cause damage to surrounding aircraft and ground equipment. Ground personnel must also be monitored closely to maintain proper distances from operating engines. Follow aircraft maintenance manual instructions and local safety protocols when running engines. Ground personnel must use proper PPE, including ear protection.

2. **What are the risks associated with performing maintenance on a turbine engine?** (AM.III.B.R2)

    Improper engine maintenance can lead to expensive repairs and damage to other parts of the aircraft. Inspections and maintenance after engine runs must be done with care due to high temperatures of the engine and components. Use proper PPE and follow all maintenance manual instructions and procedures.

3. **Describe the actions to be taken to minimize damage that might occur from an engine fire.** (AM.III.B.R3)

   Move the fuel shutoff lever to the off position if an engine fire occurs, or if the fire warning light is illuminated during the starting cycle. Continue cranking or motoring the engine until the fire has been expelled from the engine. If the fire persists, CO_2 can be discharged into the inlet duct while it is being cranked. Do not discharge CO_2 directly into the engine exhaust, because it may damage the engine. If the fire cannot be extinguished, secure all switches, and leave the aircraft. If the fire is on the ground under the engine overboard drain, discharge the CO_2 on the ground rather than on the engine. This also is true if the fire is at the tailpipe and the fuel is dripping to the ground and burning.

4. **Describe ways to decrease risk of damage due to foreign object damage (FOD).** (AM.III.B.R4)

   Inspect inlet screens on turboprop and turboshaft engines to ensure that they are not damaged. Care should be taken when working on an engine to not allow hardware, tools, or safety wire to remain in the inlet area of an engine. Before starting an engine, inspect the air inlet and the ground in front of and below the engine for objects and debris that could be injected into the engine.

Skills

1. **Identify the characteristics of different turbine compressors.** (AM.III.B.S1)

2. **Identify different types of compressors provided by the examiner and explain their operation.** (AM.III.B.S1)

3. **Given one or more turbine blades and the appropriate maintenance manual, demonstrate the correct way to measure the blades.** (AM.III.B.S2)

4. **Identify the types of turbine blades provided by the examiner.** (AM.III.B.S2)

5. **Identify the major components of turbine engines.** (AM.III.B.S3)

6. **Demonstrate the correct way to perform a prestart inspection on a turbine engine designated by the examiner.** (AM.III.B.S3)

7. **Demonstrate to the examiner the proper way to inspect a turbine engine installation to ensure that the engine and its accessories are installed in accordance with the manufacturer's specifications.** (AM.III.B.S3)

8. **Identify the airflow direction and pressure changes in turbojet engines.** (AM.III.B.S4)

9. **Demonstrate the correct way to remove, install, and connect a fuel nozzle in a turbine engine specified by the examiner.** (AM.III.B.S5)

10. **Remove and install a combustion case and liner.** (AM.III.B.S6)

11. **Inspect a combustion liner.** (AM.III.B.S6)

12. **Locate the procedures for the adjustment of a fuel control unit.** (AM.III.B.S7)

13. **Perform a turbine engine inlet guide vane and compressor blade inspection.** (AM.III.B.S8)

14. **Identify damaged inlet nozzle guide vanes.** (AM.III.B.S8)

15. **Locate the procedures for trimming a turbine engine.** (AM.III.B.S9)

16. **Demonstrate to the examiner where to find the procedure for trimming a turbine engine and explain the process.** (AM.III.B.S10)

17. **Demonstrate the correct way to use a borescope to examine the condition of a turbine blade in an engine specified by the examiner.** (AM.III.B.S11)

18. **Repair a turbine blade by blending out the damaged area. Check with the appropriate documentation to determine the maximum amount of the blade material that may be removed.** (AM.III.B.S11)

19. **Remove and/or install a turbine rotor disc.** (AM.III.B.S11)

20. **Identify the causes for turbine engine performance loss.** (AM.III.B.S12)

21. **Demonstrate the correct way to remove, check, and replace an engine bleed air valve on the engine specified by the examiner.** (AM.III.B.S12)

22. **Demonstrate to the examiner the correct way to examine the compressor section of a turbine engine for evidence of foreign object damage (FOD).** (AM.III.B.S13)

C. Engine Inspection

References: 14 CFR Parts 43, 91; AC 43.13-1; FAA-H-8083-32

Knowledge

1. **Where can an aircraft mechanic find the scope and details of the engine items that must be inspected during an annual or 100-hour inspection?** (AM.III.C.K1)
   14 CFR Part 43, Appendix D.

2. **Who is responsible for maintaining an aircraft in an airworthy condition?** (AM.III.C.K1)
   The operator or owner of the aircraft.

3. **What regulations specify the inspections that must be accomplished on a general aviation aircraft?** (AM.III.C.K1)
   14 CFR Part 91, Subpart E.

4. **Where can you find the model of magneto that is approved for an engine specified by the examiner?** (AM.III.C.K2)
   In the Type Certificate Data Sheet for the engine.

5. **Where can a list of life-limited parts be found for a specific engine?** (AM.III.C.K2)
   In the Airworthiness Limitations sections of the engine maintenance manual.

6. **Where can special inspection requirements be found for a specific engine?** (AM.III.C.K3)
   In the engine manufacturer's engine maintenance manual.

7. **What is a Supplemental Type Certificate?** (AM.III.C.K4)
   An approval issued by the FAA for a modification to a type-certificated airframe, engine, or propeller.

8. **When must FAA-approved data be used for an engine repair or modification?** (AM.III.C.K4)
   When performing a major repair or modification to an engine.

9. **Is a manufacturer's mandatory service bulletin required to be complied with?** (AM.III.C.K5)
   No, but it is recommended.

10. **Which of the following are FAA-approved data: service letters, service bulletins, instructions for continued airworthiness, ADs, or TCDSs?** (AM.III.C.K5)
    ADs and TCDSs are approved by the FAA.

11. **When must ADs be complied with?** (AM.III.C.K5)
    All ADs that apply to the affected product must be complied with in accordance with the instructions of the AD and in the timeframe specified by the AD.

12. **What paperwork is necessary for the installation of a magneto that was not approved for an engine when it was certificated, but is approved for the engine when installed according to an STC?** (AM.III.C.K6)
    A form FAA 337 must be completed stating that the installation was done in accordance with the specific STC.

13. **Where can an aircraft mechanic find the recordkeeping requirements for maintenance operations?** (AM.III.C.K6)
    14 CFR Part 43.

14. **Where can a mechanic find inspection, checking, and servicing requirements for a specific engine component?** (AM.III.C.K7)
    In the component maintenance manual (CMM) for the specific item. If the component is manufactured by the engine manufacturer, the requirements will be found in the engine maintenance manual.

15. **Where can the requirements be found for inspection and checking of engine mounts and mounting hardware?** (AM.III.C.K8)

    Engine mounts and hardware are typically airframe items, and the requirements will be found in the aircraft maintenance manual (not the engine maintenance manual).

Risk Management

1. **What cautions should be taken when performing a compression check on a reciprocating engine?** (AM.III.C.R1)

   Connecting air pressure to the cylinder can cause the propeller to suddenly rotate with force. Carefully contain the propeller prior to connecting air to the cylinder. This task is best accomplished with two people: one controlling the propeller and the other controlling the air pressure.

2. **What are the risks associated with performing maintenance adjustments on an operating reciprocating engine?** (AM.III.C.R2)

   The danger of performing maintenance on an operating engine is the close proximity to the turning propeller that, if encountered, will cause injury or death. While the engine is operating, determine what adjustments need to be made. Turn the engine off, make the adjustments, and then start the engine back up to verify the adjustments.

3. **What are the risks associated with performing maintenance on an operating turbine engine?** (AM.III.C.R3)

   If the turbine engine is a turboprop, the spinning propeller represents the danger. On a turbojet or turbofan engine, the dangers are the engine inlet and the exhaust. Follow maintenance manual instructions for making any adjustments to a turbine engine.

Skills

1. **Perform a differential compression test on an engine specified by the examiner. Explain to the examiner what could cause low compression on one or more cylinders, and what action could be taken to correct the situation.** (AM.III.C.S1)

2. **Explain to the examiner the changes in an inspection program caused by a modification that was done according to an airworthiness directive.** (AM.III.C.S2)

3. **Evaluate the assigned engine for configuration in accordance with FAA-approved or manufacturer data.** (AM.III.C.S2)

4. **Inspect an aircraft engine for service bulletin compliance.** (AM.III.C.S3)

5. **Determine if an aircraft engine's maintenance manual is current.** (AM.III.C.S3)

6. **Demonstrate to the examiner how to evaluate the powerplant maintenance records for compliance with requirements.** (AM.III.C.S3)

7. **Demonstrate to the examiner how to inspect for compliance with applicable ADs.** (AM.III.C.S4)

8. **Determine engine installation eligibility for the assigned configuration.** (AM.III.C.S5)
9. **Determine engine conformity with engine specifications or type certificate data sheet.** (AM.III.C.S5)
10. **Identify an engine type without reference material other than the data plate.** (AM.III.C.S6)
11. **Identify the various components installed on an engine.** (AM.III.C.S6)
12. **Determine if the assigned engine conforms to engine specifications and the engine TCDS.** (AM.III.C.S6)
13. **Perform an over temperature inspection.** (AM.III.C.S7)
14. **Perform an engine over torque inspection.** (AM.III.C.S7)
15. **Perform an aircraft engine over speed inspection.** (AM.III.C.S7)
16. **Determine the conformity of installed spark plugs or igniters.** (AM.III.C.S7)
17. **Demonstrate how to perform the assigned inspection on an engine.** (AM.III.C.S7)
18. **Check the engine controls for freedom of operation.** (AM.III.C.S8)
19. **Inspect engine controls for proper adjustment.** (AM.III.C.S8)
20. **Inspect an engine for fluid leaks after performing maintenance.** (AM.III.C.S9)
21. **Inspect an aircraft engine's accessories for conformity.** (AM.III.C.S10)
22. **Inspect an aircraft engine accessory for serviceability.** (AM.III.C.S10)
23. **Check the records of an engine specified by the examiner for a list of all applicable airworthiness directives. Explain to the examiner the way a specific AD should be complied with, and write up a maintenance record entry for the compliance.** (AM.III.C.S11)
24. **Inspect an aircraft turbine engine for records time left on any life-limited parts.** (AM.III.C.S11)
25. **Demonstrate the correct way to determine the allowable cycle life of a turbine engine component specified by the examiner.** (AM.III.C.S11)
26. **Given the operating records of a turbine engine, demonstrate the way to calculate the cycle life between overhauls of a component specified by the examiner.** (AM.III.C.S11)

27. **Demonstrate to the examiner the correct way to start a turbine engine. Explain the way to detect a hot start and a hung start, and the correct procedure to follow if either occurs.** (AM.III.C.S12)

28. **Demonstrate to the examiner the correct way to start and operate a reciprocating engine.** (AM.III.C.S12)

29. **Conduct a 100-hour inspection on an aircraft engine specified by the examiner using the checklist prepared by the aircraft manufacturer.** (AM.III.C.S13)

30. **Inspect the assigned engine mounts.** (AM.III.C.S14)

31. **Inspect an engine mount to determine serviceability.** (AM.III.C.S14)

D. Engine Instrument Systems

References: AC 43.13-1; FAA-H-8083-32

Knowledge

1. **What are three types of flow-indicating systems used for turbine engines?** (AM.III.D.K1)

   Vane-type flow meters, synchronous mass-type flow meters, and electronic motorless mass flow meters.

2. **What do mass flow systems measure?** (AM.III.D.K1)

   The mass of the flow. This is affected by the density of the fuel which is in turn affected by the fuel temperature.

3. **In what units is an electronic motorless mass flow meter system calibrated?** (AM.III.D.K1)

   Pounds per hour of fuel flow.

4. **What is indicated in a turbine engine if the fuel flow for all conditions is high?** (AM.III.D.K1)

   If there are other instrument abnormalities, there are possibly damaged turbine components.

5. **Who is authorized to repair a fuel flow indicating system?** (AM.III.D.K1)

   The components can be repaired only by an FAA approved repair station certificated for the particular instruments.

6. **Where can you find the electrical power requirements for a fuel flow indicating system?** (AM.III.D.K1)

   In the aircraft maintenance manual.

7. **Where in the fuel system is the fuel flow transmitter located?** (AM.III.D.K1)
   Between the fuel control and the fuel nozzles.

8. **What kind of indicating system is used to measure oil temperature in a small, single-engine, general aviation airplane?** (AM.III.D.K2)
   A sealed system in which a bulb, partially filled with a highly volatile liquid, is installed inside the oil strainer. The bulb is connected to a pressure gauge by a copper tube. The pressure of the gas above the liquid is proportional to the temperature of the oil that surrounds the bulb, and the pressure gauge is calibrated in terms of temperature.

9. **What type of instrument system is used for measuring cylinder-head temperature and exhaust gas temperature?** (AM.III.D.K2)
   A thermocouple system.

10. **What kind of electrical instrument system is used for measuring oil temperature?** (AM.III.D.K2)
    A ratiometer-type system.

11. **What should a thermocouple cylinder-head temperature indicator read when the engine is cold?** (AM.III.D.K2)
    The same as the outside air temperature gauge.

12. **What is critical in the installation of the thermocouple leads for a cylinder-head temperature indicator, the length of the leads or their resistance?** (AM.III.D.K2)
    The resistance is critical, it must have the resistance specified on the indicator. This is usually 2 or 8 ohms.

13. **What should a ratiometer temperature indicator read when the electrical power is off?** (AM.III.D.K2)
    The pointer should be off scale on the low side.

14. **How does a ratiometer-type oil temperature gauge measure the temperature of the oil?** (AM.III.D.K2)
    A ratiometer measures the resistance of the temperature bulb. The indicator is calibrated in degrees Fahrenheit or Celsius rather than in ohms.

15. **What kind of tachometer is used on most small single-engine airplanes?** (AM.III.D.K3)
    A magnetic drag tachometer that is similar to an automobile speedometer.

16. **What would likely cause the pointer of a magnetic drag tachometer to oscillate?** (AM.III.D.K3)
    Either a dry cable connecting the indicator to the engine, or a kink in the cable housing.

17. **What kind of tachometer is used on most large multi-engine airplanes?** (AM.III.D.K3)
    A three-phase AC electric tachometer.

18. What is measured to indicate the engine RPM with a three-phase AC electric tachometer? (AM.III.D.K3)

The frequency of the AC the tachometer generator produces.

19. What kind of indicator mechanism is used to measure oil pressure? (AM.III.D.K4)

A Bourdon tube.

20. What kind of indicator mechanism is used to measure manifold pressure? (AM.III.D.K4)

A differential bellows.

21. What should a manifold pressure gauge read when the engine is not operating? (AM.III.D.K4)

The existing barometric pressure as indicated on the altimeter barometric scale when the altimeter pointers are set to the surveyed field elevation.

22. What type of fitting is used to connect an oil pressure gauge to the engine? (AM.III.D.K4)

A fitting with a restrictor that smooths out the pulsations and should a break occur in the indicator line, prevents a serious loss of engine oil.

23. What is used to fill the line between a fuel pressure gauge and the fuel metering system? (AM.III.D.K4)

A light oil such as kerosine.

24. What could cause a fuel pressure gauge to fluctuate? (AM.III.D.K4)

The loss of the light oil in the line between the gauge and the engine.

25. What is the range of pressure needed for a fuel pressure gauge used with a float carburetor? (AM.III.D.K4)

Normally from 1 to 25 psi.

26. What are the colors used in annunciator indicating systems, and what do they represent? (AM.III.D.K5)

Green indicates a safe condition, yellow indicates a caution alert, and red indicates a warning alert.

27. What does a torquemeter measure? (AM.III.D.K6)

Most torque systems use an oil pressure output from a torque valve to indicate actual engine power output at various power settings. The torquemeter indicates the amount of torque being produced at the propeller shaft. A helical gear moves back and forth as the torque on the propeller shaft varies. This gear, acting on a piston, positions a valve that meters the oil pressure proportionally to the torque being produced.

28. What does the engine pressure ratio (EPR) indicator measure? (AM.III.D.K7)

Engine pressure ratio (EPR) is an indication of the thrust being developed by a turbofan engine and is used to set power for takeoff on many types of aircraft. It is instrumented by total pressure pickups in the engine inlet and in the turbine exhaust.

29. **What is an EICAS (engine indicating and crew alerting system)?** (AM.III.D.K8)
ECIAS is an electronic instrumentation system that monitors airframe and engine parameters and displays the essential information on a video display on the instrument panel. Only vital information is continually displayed, but when any sensed parameters fall outside of their allowable range of operation, they are automatically displayed.

30. **What is an engine FADEC system?** (AM.III.D.K9)
FADEC (full-authority digital electronic control) is a digital electronic fuel control for a gas turbine engine that is functioning during all engine operations, hence full authority. It includes the EEC (electronic engine control) and functions with the flight management computer. FADEC schedules the fuel to the nozzles in such a way that prevents overshooting power changes and over-temperature conditions. FADEC furnishes information to the EICAS (engine indicating and crew alerting system).

31. **What is an ECAM system?** (AM.III.D.K10)
ECAM (electronic centralized aircraft monitor) is a system that monitors airframe and engine functions and displays them in the flight deck. It can also provide checklists for system failures.

32. **What do the colors red, green, and yellow indicate on an aircraft instrument or gauge?** (AM.III.D.K11)
Red indicates a maximum range or limit, green indicates a normal operating range, and yellow indicates a caution range.

Risk Management

1. **What risks are associated with inadvertent damage to an instrument or indicating system during maintenance operations?** (AM.III.D.R1)
The dangers arise from incorrect readings on flight deck instruments, with potential operations of the aircraft outside of design limits. Care must be taken not to damage indicating systems or instruments.

2. **What are the dangers of instrument error due to failures or out-of-calibration conditions?** (AM.III.D.R2)
Instruments that are out of calibration or have faulty indication can cause actions to be taken (or not taken) that are outside of operating parameters. Instruments are to be calibrated according to regulations and manufacturer's recommendations.

Skills

1. **Troubleshoot an engine oil temperature/pressure instrument system.** (AM.III.D.S1)

2. **Explain to the examiner the troubleshooting procedure for an oscillating cylinder-head temperature indicator.** (AM.III.D.S1)

3. **Demonstrate to the examiner the way to replace a thermocouple probe for a cylinder-head temperature indicating system.** (AM.III.D.S1)

4. **Troubleshoot a low fuel pressure indicating system.** (AM.III.D.S2)

5. **Remove, inspect, and install a fuel-flow transmitter.** (AM.III.D.S3)

6. **Remove, inspect, and install a fuel flow gauge.** (AM.III.D.S4)

7. **Locate and identify the tachometer generator on an engine specified by the examiner.** (AM.III.D.S5)

8. **Check the fuel-flow transmitter power supply voltage on the supplied system.** (AM.III.D.S6)

9. **Inspect a tachometer's markings for accuracy.** (AM.III.D.S7)

10. **Demonstrate to the examiner the way to check all the components in a cylinder-head temperature indicating system.** (AM.III.D.S8)

11. **Perform resistance measurements of a thermocouple indication system.** (AM.III.D.S8)

12. **Remove, inspect, and/or install a turbine engine exhaust gas temperature (EGT) harness.** (AM.III.D.S9)

13. **Locate the procedures for troubleshooting a turbine engine pressure ratio (EPR) system.** (AM.III.D.S10)

14. **Demonstrate to the examiner the way to check an electric tachometer system for a failure of the system to indicate.** (AM.III.D.S11)

15. **Demonstrate the way to check or replace the tachometer cable for a magnetic drag tachometer.** (AM.III.D.S11)

16. **Replace a cylinder head temperature thermocouple.** (AM.III.D.S12)

17. **Inspect a cylinder head temperature indicating system.** (AM.III.D.S12)

18. **Replace a probe in an exhaust gas temperature system specified by the examiner.** (AM.III.D.S13)

19. **Demonstrate to the examiner how to inspect EGT probes.** (AM.III.D.S13)

20. **Troubleshoot a fuel-flow system.** (AM.III.D.S14)

21. **Locate and inspect an engine's low fuel pressure warning system components.** (AM.III.D.S14)

22. **Check an aircraft engine manifold pressure gauge for proper operation.** (AM.III.D.S15)

23. **Check for proper operation of a manifold pressure gauge.** (AM.III.D.S15)

24. **Inspect a leaking manifold pressure system.** (AM.III.D.S16)

25. **Repair a low oil pressure warning system.** (AM.III.D.S17)

26. **Troubleshoot an EGT indicating system.** (AM.III.D.S18)

27. **Locate and identify the oil temperature bulb on an engine specified by the examiner. Explain the precautions to be taken when replacing the bulb.** (AM.III.D.S19)

28. **Demonstrate to the examiner the way to check the capillary tube that connects a mechanical oil temperature gauge to the engine.** (AM.III.D.S19)

E. Engine Fire Protection Systems

References: AC 43.13-1; FAA-H-8083-32

Knowledge

1. **What are the engine fire zones?** (AM.III.E.K1)

   The powerplant installation has several designated fire zones, which are:
   1. the engine power section;
   2. the engine accessory section;
   3. except for reciprocating engines, any complete powerplant compartment in which no isolation is provided between the engine power section and the engine accessory section;
   4. any APU compartment;
   5. any fuel-burning heater and other combustion equipment installation;
   6. the compressor and accessory sections of turbine engines; and
   7. combustor, turbine, and tailpipe sections of turbine engine installations that contain lines or components carrying flammable fluids or gases.

2. **What different types of fires can occur in an engine compartment?** (AM.III.E.K1)

   On reciprocating engines, fires can occur from fuel leaks spraying on hot engine components, exhaust leaks streaming hot gases on other components or the engine cowling, and electrical faults. Turbine engine failures can cause turbine blades to melt and separate from the engine, causing damage to the exhaust duct and surrounding aircraft structure or fuel lines. A fuel nozzle that is flowing too much fuel can cause burn-through of the combustion liner or tail cone.

3. **What are four types of engine fire detection systems?** (AM.III.E.K2)

   Thermoswitch system, thermocouple system, thermistor-type continuous-loop system, pneumatic-type continuous-loop system.

4. **Which type of fire detection system operates on the principle of the rate of temperature rise?** (AM.III.E.K2)

   The thermocouple system.

5. **Why must the reference junction in a thermocouple fire detection system be thermally insulated?** (AM.III.E.K2)

   This is a rate-of-temperature-rise system. The reference junction must be insulated so its temperature will not rise as quickly as that of the measuring junctions if a fire should occur.

6. **How do you check a thermocouple fire detection system for operation?** (AM.III.E.K2)

   Press the test switch. This heats the test thermocouple enough for it to close the sensitive relay, which in turn closes the slave relay and activates the fire warning light and bell.

7. **How is a pneumatic-type continuous-loop fire detection system checked for operation?** (AM.III.E.K2)

   When the test switch is closed, low-voltage AC flows through the tube that encloses the sensitive element. This current heats the element enough for it to release sufficient gas to activate the alarm switch.

8. **What would likely cause a false fire alarm in a thermistor-type continuous-loop system?** (AM.III.E.K3)

   A kinked or pinched continuous-loop element.

9. **If there is a break in a single-wire continuous-loop fire detector element, and the system tests bad, will it indicate the presence of a fire?** (AM.III.E.K3)

   Yes, even though it tests bad, it will still indicate the presence of a fire.

10. **What maintenance is allowed on a continuous-loop fire detection system?** (AM.III.E.K3)

    Inspect and replace as needed the continuous-loop element, the controller, and the fire warning bell and/or light.

11. **What is used as a fire extinguishing agent in some older aircraft?** (AM.III.E.K4)

    Carbon dioxide (CO_2).

12. **What fire extinguishing agent is used in the high-rate discharge bottles installed in a jet transport aircraft?** (AM.III.E.K4)

    Halon 1211 or 1301.

13. **How is a high-rate discharge bottle of fire extinguishing agent discharged?** (AM.III.E.K4)

    A powder charge is ignited and it blows a cutter through a frangible disc that seals the bottle.

14. **What precaution must be taken when checking the electrical circuit for igniting the powder charge in the valve of a high-rate discharge bottle?** (AM.III.E.K4)

    The current used to check the integrity of the electric circuit must not be high enough to ignite the powder charge.

15. **What color and marking identifies the fluid lines that carry the fire extinguishing agent in an aircraft?** (AM.III.E.K4)

    Brown tape with a series of diamonds on it.

16. **What happens in a jet transport airplane when the fire-pull T-handle for an engine is pulled?** (AM.III.E.K5)

    a. The engine fuel shutoff valve is closed, and the generator field relay is tripped.
    b. The compressor bleed air valve is closed.
    c. The anti-ice valve is closed.
    d. The hydraulic supply shutoff valve is closed.
    e. The hydraulic pump low-pressure warning lights are turned off.

17. **What is indicated if the red discharge indicator for a fire extinguishing system is blown out?** (AM.III.E.K5)

    The system has been discharged by an overheat condition.

18. **What is indicated if the yellow discharge indicator for a fire extinguishing system is blown out?** (AM.III.E.K5)

    The system has been discharged by actuating the bottle discharge switch.

Risk Management

1. **What are the dangers associated with the handling and storage of discharge cartridges?** (AM.III.E.R1)

   A discharge cartridge, or squib, contains an explosive charge that must be stored and handled in accordance with the maintenance manual instructions.

2. **What are the risks associated with extinguishing agents used in aircraft systems?** (AM.III.E.R2)

   Many of these systems are high-pressure systems that could cause injury to maintenance personnel. In addition, the chemicals used are considered hazardous materials and must be handled with care. Follow aircraft maintenance manual cautions and procedures when performing maintenance on, or near, fire extinguishing systems.

3. **What precautions should be taken when working on fire extinguisher discharge cartridges (squibs)?** (AM.III.E.R3)

   An electric current from the fire extinguishing circuit fires the squib. Improper procedures could inadvertently fire the explosive charge in the squib. Always follow the maintenance manual procedures when disconnecting or connecting a squib.

Skills

1. **Troubleshoot and repair an engine fire detection system.** (AM.III.E.S1)

2. **Identify to the examiner the valves that are shut off when the fire-pull T-handle is pulled.** (AM.III.E.S2)

3. **Identify the fire detection sensing units.** (AM.III.E.S2)

4. **Check and/or inspect a fire detection warning system.** (AM.III.E.S2)

5. **Locate the troubleshooting information for a fire detection system.** (AM.III.E.S2)

6. **Demonstrate to the examiner the correct way to check a continuous-loop fire detection system for integrity.** (AM.III.E.S3)

7. **Demonstrate the correct way to attach a section of continuous-loop fire detection element to the aircraft structure.** (AM.III.E.S3)

8. **Inspect the fire detection continuous loop system.** (AM.III.E.S3)

9. **Inspect a fire detection thermal switch or thermocouple system.** (AM.III.E.S4)

10. **Demonstrate to the examiner where to find the troubleshooting procedures for a fire extinguisher system.** (AM.III.E.S5)

11. **Inspect an engine's fire extinguisher system blowout plugs.** (AM.III.E.S6)

12. **Inspect a turbine engine fire detection system.** (AM.III.E.S7)

13. **Inspect a fire extinguisher discharge circuit.** (AM.III.E.S8)

14. **Troubleshoot a fire protection system.** (AM.III.E.S9)

15. **Check the operation of firewall shutoff valves.** (AM.III.E.S9)

16. **Troubleshoot and repair a fire extinguishing system.** (AM.III.E.S9)

17. **Demonstrate to the examiner the correct way to determine the state of charge of a CO_2 fire extinguisher bottle.** (AM.III.E.S10)

18. **Demonstrate to the examiner the correct way to determine the state of charge of an HRD fire extinguisher bottle.** (AM.III.E.S10)

19. **Demonstrate to the examiner the way to check the squib in a high-rate-discharge bottle of fire extinguishing agent for the shelf-life date.** (AM.III.E.S10)

20. **Check the fire extinguisher discharge circuit.** (AM.III.E.S10)

21. **Inspect a fire extinguisher system for hydrostatic test requirements.** (AM.III.E.S11)

22. **Check the flame detectors for operation.** (AM.III.E.S12)

23. **Check operation of fire warning press-to-test and troubleshoot faults.** (AM.III.E.S13)

24. **Identify the components of the continuous-loop fire detection system.** (AM.III.E.S14)

F. Engine Electrical Systems

References: AC 43.13-1; FAA-H-8083-30, FAA-H-8083-32

Knowledge

1. **What is used as the rectifier to produce direct current in a DC generator?** (AM.III.F.K1)
   Brushes and a commutator.

2. **What is used as the rectifier to produce direct current in a DC alternator?** (AM.III.F.K2)
   Six solid-state diodes.

3. **Why is it not necessary to flash the field of a DC alternator after it has been overhauled?** (AM.III.F.K2)
   An alternator field is excited by battery current and residual voltage is not used to start the alternator producing current.

4. **When should the brushes in a starter motor be replaced?** (AM.III.F.K3)
   When they have worn to one half of their original length.

5. **Is a starter motor series-wound or shunt-wound?** (AM.III.F.K3)
   Series wound for maximum stalled-rotor torque.

6. **What is a starter-generator?** (AM.III.F.K3)
   A single, engine-mounted component that serves as a starter for starting the turbine engine. When the engine is running, the circuitry automatically shifts so it acts as a compound-wound generator.

7. **If the installed ammeter gauge is redlined at 100 percent of generator or alternator rating, and the regulator does not control current, how is maximum current controlled?** (AM.III.F.K4)
   Current is controlled by the pilot or operator of the aircraft by not applying loads greater than redline, except for short, intermittent loads.

8. **If placards or monitoring devices are not practical or desired, what practice protects the battery and electrical system?** (AM.III.F.K4)
   The generator is sized so that total continuous load of the electrical system does not exceed 80 percent of the rated capacity of the generator or alternator.

9. **When determining maximum continuous electrical load of an aircraft, what defines if the load is intermittent or continuous?** (AM.III.F.K4)
   The length of time the system is used. If the component or system is used for less than two minutes, it is considered an intermittent load. If it is used for more than two minutes, it is considered a continuous load.

10. **How is generator output voltage controlled?** (AM.III.F.K4)
    By controlling the field current strength through the use of a voltage regulator.

11. **How is the direction of rotation of a DC electric motor reversed?** (AM.III.F.K5)
    Reverse the polarity of the armature or the field, but not both.

12. **What must be done to a DC generator, after it has been overhauled, before it can produce electricity?** (AM.III.F.K5)
    The field must be flashed to restore residual magnetism to the field frame so it can begin to produce current.

13. **How many phases of AC electricity are produced by an alternator?** (AM.III.F.K6)
    Three phases.

14. **How many diodes are required to convert three phase AC to DC electricity?** (AM.III.F.K6)
    Six diodes.

15. **What factors must be considered when determining wire size?** (AM.III.F.K7)
    a. System voltage
    b. If the circuit has an intermittent or continuous load
    c. Current
    d. Distance in feet of the wire run
    e. The temperature rating of the wire to be used
    f. The ambient temperature of the area where the wire will be installed
    g. The number of wires in the wire bundle
    h. The percentage of time the wire bundle will be loaded
    i. The maximum altitude the wire will see

16. **When paralleling the outputs of two generators or alternators, how close must the output voltages be?** (AM.III.F.K8)
    Within a few tenths of a volt. Follow the instructions in the maintenance manual for specified voltages.

17. **Why is it important to parallel the outputs of a dual-generator system?** (AM.III.F.K8)
    If the generator output voltages are not the same, one generator will attempt to carry all, or most, of the load, causing the generator to become overloaded.

18. **What is the purpose of a generator CSD (constant speed drive) unit?** (AM.III.F.K9)
    A CSD maintains the output frequency of a AC generator by maintaining the speed of the generator through varying engine speeds.

19. **What is an IDG (integrated drive unit)?** (AM.III.F.K9)
    An IDG combines the AC generator and CSD into a single unit.

20. **What are two ways wires can be attached to the pins in a cannon plug?** (AM.III.F.K10)
    a. By soldering the wires into pots on the pins and sockets.
    b. By crimping tapered pins onto the wires and inserting the pins into tapered holes in the pins and sockets.

21. **Why is stranded wire rather than solid wire used in most powerplant electrical systems?** (AM.III.F.K10)

    Solid wire is likely to break when it is subjected to vibration.

22. **What two things must be considered in selection of wire size when making an electrical installation in an aircraft?** (AM.III.F.K10)

    The current-carrying capability of the wire, and the voltage drop caused by current flowing through the wire.

23. **What is used to protect a wire bundle from chafing where it passes through a hole in a bulkhead or frame?** (AM.III.F.K10)

    A grommet around the edges of the hole.

24. **How are electrical wires protected where they pass through an area of high temperature?** (AM.III.F.K10)

    Wires passing through these areas are insulated with high-temperature insulation, and the wires are enclosed in some type of protective conduit.

25. **What is the minimum separation allowed between a wire bundle and a fluid line that carries combustible fluid or oxygen?** (AM.III.F.K10)

    Six inches whenever possible. If not practical, the wires should not run parallel to fuel or oxygen lines and a minimum of two inches must be maintained.

26. **Which aircraft electrical circuit does not normally contain a fuse or circuit breaker?** (AM.III.F.K10)

    The starter motor circuit.

27. **When should aircraft wiring be installed in a conduit?** (AM.III.F.K10)

    When the wiring passes through an area in the aircraft where open wiring could likely be damaged, such as through a wheel well.

28. **What is meant by a trip-free circuit breaker?** (AM.III.F.K10)

    A circuit breaker that opens a circuit any time an excessive amount of current flows, regardless of the position of the circuit breaker's operating handle.

29. **Which way should the toggle of a switch that controls the propeller pitch move to place the propeller in low pitch (high RPM)?** (AM.III.F.K10)

    Forward.

30. **Why are protective covers placed over some switches in an aircraft electrical circuit?** (AM.III.F.K10)

    To prevent the switch from being inadvertently actuated.

Risk Management

1. **Why is important to maintain proper polarity on components or systems when performing maintenance on electrical systems?** (AM.III.F.R1)

   Failure to maintain proper polarity can quickly cause damage to components and aircraft wiring. Care must be taken to observe and follow proper polarity.

2. **What are the risks associated with not understanding the meaning of warning or caution lights on an annunciator panel?** (AM.III.F.R2)

   Not knowing or understanding the meaning of warning and caution lights can result in incorrect operation of aircraft engines and systems, faulty troubleshooting procedures, and improper repairs.

3. **What cautions should be taken when performing maintenance on energized aircraft circuits and systems?** (AM.III.F.R3)

   An energized system can move if someone moves a flight deck control. This is more of a challenge on large aircraft where individuals working on a system cannot be seen from the flight deck. Circuit breakers should be pulled and tagged when working in an area that could present a danger to personnel. Dangerous engine components include thrust reversers, which move quickly and with much force.

4. **What risks are associated with routing wiring near flammable fluid lines?** (AM.III.F.R4)

   The danger is that lines can leak, flammable fluid can drip on wires, and fumes can be created in closed spaces. If wiring is not secured correctly, it can chafe on fuel lines or nearby structure, causing sparks and potential fire. Wiring should be routed above flammable fluid lines and secured in a manner that prevents chafing.

Skills

1. **Demonstrate to the examiner how to inspect engine electrical wiring, switches, and protective devices.** (AM.III.F.S1)

2. **Use publications to determine replacement part numbers.** (AM.III.F.S2)

3. **Replace an engine-driven generator or alternator.** (AM.III.F.S3)

4. **Troubleshoot a voltage regulator in an aircraft electrical generating system.** (AM.III.F.S3)

5. **Install a generator on an engine specified by the examiner. Check the voltage output to determine whether it meets the manufacturer's specifications for voltage at a specified RPM.** (AM.III.F.S3)

6. **Parallel a dual-generator electrical system.** (AM.III.F.S4)

7. **Inspect an engine-driven generator or alternator.** (AM.III.F.S4)

8. **Demonstrate to the examiner how to troubleshoot an aircraft electrical generating system.** (AM.III.F.S5)

9. **Repair an engine direct-drive electric starter.** (AM.III.F.S6)

10. **Visually identify and describe the operation of components in a constant speed drive (CSD) or integrated drive generator (IDG).** (AM.III.F.S6)

11. **Troubleshoot a direct-drive electric starter system.** (AM.III.F.S7)

12. **Demonstrate to the examiner how to inspect an electrical system cable.** (AM.III.F.S8)

13. **Determine the wire size for an engine electrical system.** (AM.III.F.S9)

14. **Using the electrical wire chart in AC 43.13-1B, page 11-23, find the wire size needed to carry 20 amps of current for 30 feet in a 28-volt circuit.** (AM.III.F.S9)

15. **Demonstrate the correct way to splice a wire in an electrical circuit.** (AM.III.F.S10)

16. **Repair a broken engine electrical system wire.** (AM.III.F.S10)

17. **Replace or install lacing on a wire bundle.** (AM.III.F.S11)

18. **Using an electrical system schematic diagram for a starter-generator installation, explain to the examiner the way the unit acts as a starter and then shifts into the function of a generator.** (AM.III.F.S12)

19. **Install a tachometer generator and check its operation.** (AM.III.F.S12)

20. **Troubleshoot an electrical system using a schematic or wiring diagram.** (AM.III.F.S12)

21. **Fabricate a bonding jumper.** (AM.III.F.S13)

22. **Inspect a turbine engine starter generator.** (AM.III.F.S14)

23. **Inspect engine electrical connectors.** (AM.III.F.S15)

G. Engine Lubrication Systems

References: AC 43.13-1; FAA-H-8083-32

Knowledge

1. **What is straight mineral oil?** (AM.III.G.K1)
   It is the lubricating oil as obtained by fractional distillation of crude oil. It does not have any additives.

2. **What is ashless dispersant (AD) oil?** (AM.III.G.K1)
   A mineral lubricating oil containing additives that disperse the contaminants throughout the oil so they will not clump and clog oil passages. There are no ash-forming additives in AD oil.

3. **What is synthetic oil?** (AM.III.G.K1)
   A lubricating oil that is made by chemically changing the nature of an oil base to give it the needed characteristics. Synthetic oil is the primary oil for turbine engines.

4. **Is automotive oil suitable for use in aircraft engines?** (AM.III.G.K1)
   No, automotive and aviation oils are formulated for entirely different operating conditions.

5. **What is meant by a multiviscosity oil?** (AM.III.G.K1)
   It is a lubricating oil with a viscosity index improver that increases the viscosity of the oil when it is hot and decreases the viscosity when it is cold.

6. **Where can you find the grade of engine oil specified for a particular aircraft?** (AM.III.G.K1)
   In the airplane flight manual or the pilot's operating handbook for the aircraft.

7. **What are six functions of the oil in an aircraft engine?** (AM.III.G.K2)
   a. Reduces friction.
   b. Seals and cushions.
   c. Removes heat.
   d. Cleans inside the engine.
   e. Protects against corrosion.
   f. Performs hydraulic action.

8. **What is the purpose of an air-oil separator, or deaerator, in a turbine engine oil tank?** (AM.III.G.K2)
   In normal operation, the oil picks up a quantity of air and it is swirled as it enters the deaerator. The swirling action releases the air from the oil, and the air is used to pressurize the oil tank.

9. **How is oil pressure regulated in an aircraft engine?** (AM.III.G.K2)
   A pressure-relief valve senses the desired pressure and sends excess oil back into the engine sump.

10. **What is the purpose of the restricted orifice in the line between the oil pressure gauge and the engine?** (AM.III.G.K2)
    The restricted fitting helps dampen any pulsations in the oil pressure caused by the pump.

11. **What is a hot-tank lubrication system for a turbojet engine?** (AM.III.G.K2)
    A lubrication system in which the oil cooler is in the pressure subsystem and the scavenged oil is not cooled before it is returned to the tank.

12. **What is a cold-tank lubrication system for a turbojet engine?** (AM.III.G.K2)
    A lubrication system in which the oil cooler is in the scavenge subsystem, and the scavenged oil is cooled before it is returned to the tank.

13. **Where does a wet sump oil system store oil?** (AM.III.G.K3)
    In a reservoir inside the engine.

14. **Where does a dry sump oil system store oil?** (AM.III.G.K4)
    In a reservoir outside the engine through the use of a scavenge pump and external tubing.

15. **What is the purpose of a chip detector?** (AM.III.G.K5)
    Magnetic chip detectors are used in the oil system to detect and catch ferrous (magnetic) particles present in the oil. During maintenance, the chip detectors are removed from the engine and inspected for metal; if none is found, the detector is cleaned, replaced, and safety wired. If metal is found on a chip detector, an investigation should be made to find the source of the metal on the chip.

16. **What components in a reciprocating-engine lubrication system must be inspected on a 100-hour or annual inspection?** (AM.III.G.K5)
    The oil sump or tank, the oil strainer screen, the oil filter, the oil cooler and the temperature control valve, the oil pressure and temperature gauges and any transmitters associated with them, and the entire engine for indication of oil leaks.

17. **What two instruments show the condition of a reciprocating-engine lubrication system?** (AM.III.G.K6)
    Oil pressure and oil temperature gauges.

18. **What must be done to an oil filter when it is removed from the engine?** (AM.III.G.K6)
    It should be cut open and the pleated element examined to determine the type and amount of contaminant carried by the oil.

19. **What is a spectrometric oil analysis program?** (AM.III.G.K6)
    A program in which a sample of oil is taken from the engine at regular intervals and sent to a laboratory, where it is burned in an electric arc. The resulting light is analyzed for the wavelengths of the elements that are present in the oil sample. Traces of aluminum, copper, and iron in the oil indicate wear of the pistons or wrist pin plugs (aluminum), cylinder walls or piston rings (iron), or main bearings or bushings (copper). A single sample is meaningless. There must be a series of samples taken at regular intervals to measure the change in the amounts of these metals.

20. **What type of oil quantity indicator is used in most aircraft engines?** (AM.III.G.K6)
    A dipstick that measures the quantity of oil in the tank or sump.

21. **When should the engine oil quantity be checked on a turbine engine?** (AM.III.G.K6)
    As soon as practical after the engine is shut down.

22. **Where is the oil temperature measured on a reciprocating engine?** (AM.III.G.K6)
    Usually at the oil pressure screen before the oil goes into the engine passages.

23. **Why do engine manufacturers recommend that engine lubricating oil be changed at specific intervals?** (AM.III.G.K6)
    The oil picks up contaminants and carries them through the engine where they can cause wear. The oil also becomes acidic and causes corrosion in the engine.

24. **What problem can improper piston ring conditioning (also known as run-in or break-in) lead to?** (AM.III.G.K7)

    High oil consumption.

25. **What are causes of high oil consumption in a reciprocating engine?** (AM.III.G.K7)

    a. Failing or failed bearing
    b. Worn or broken piston rings
    c. Incorrect installation of piston rings
    d. External oil leakage
    e. Leakage through engine fuel pump vent
    f. Engine breather or vacuum pump breather

Risk Management

1. **What is the danger of mixing engine oils?** (AM.III.G.R1)

   Oils may not be compatible with each other and can cause the degrading of the oil properties leading to unsatisfactory lubrication of the engine. When adding oil to an engine, add the same type of oil that was used previously.

2. **What are the risks of using engine lubricants other than those specified by the manufacturer?** (AM.III.G.R2)

   The engine was tested and certified with specific lubricants that allowed the engine manufacturer to meet the regulations for the engine design. The use of lubricants that were not certified may lead to dangerous wear and lubrication issues in the engine, leading to premature failure. Follow the manufacturer's recommendations when using engine lubricants.

3. **What cautions should be used when handling, storing, and disposing of used lubricating oils?** (AM.III.G.R3)

   Used oils are a hazardous material that is damaging to the environment. Proper handling, storage, disposal, and recycling methods must be followed carefully.

Skills

1. **Service the lubrication system in a turbine engine specified by the examiner. Choose the correct oil and explain the way the oil quantity is indicated.** (AM.III.G.S1)

2. **Using the proper documentation for an engine specified by the examiner, locate the part number for the oil cooler. Describe the proper way to remove and replace the cooler and prepare the engine for return to service including testing for leaks.** (AM.III.G.S1)

3. **Inspect an oil cooler and oil lines.** (AM.III.G.S1)

4. **Using the Type Certificate Data Sheets for an aircraft specified by the examiner, find the oil quantity specified and the amount of undrainable oil that is trapped in the system.** (AM.III.G.S2)

5. **Demonstrate to the examiner where to find the turbine engine oil filter bypass indicator on the assigned system.** (AM.III.G.S3)

6. **Using a list of oils approved for an aircraft engine, choose the correct oil for the existing climatic conditions and explain to the examiner the reason for your choice.** (AM.III.G.S4)

7. **Demonstrate the way to take a sample of oil for a spectrometric oil analysis.** (AM.III.G.S5)

8. **Demonstrate to the examiner the proper way to change the oil filter on an aircraft engine. Inspect the used filter for contamination and explain to the examiner the source of the contaminants. Explain the precautions that must be taken to avoid damaging the engine when installing the new filter.** (AM.III.G.S6)

9. **Demonstrate the correct way to adjust the oil pressure on an engine specified by the examiner.** (AM.III.G.S7)

10. **Demonstrate the correct way to preoil an engine specified by the examiner.** (AM.III.G.S7)

11. **Using a diagram of a turbine-engine lubrication system, identify the filters, spray nozzles, pumps, relief valves, check valves, and bypass valves.** (AM.III.G.S8)

12. **Identify oil system components on the assigned engine.** (AM.III.G.S8)

13. **Replace a gasket or seal in the lubrication system of an engine specified by the examiner. Run the engine and check the entire system for leaks. Correct any leaks found.** (AM.III.G.S9)

14. **Replace an oil system component assigned by the examiner.** (AM.III.G.S9)

15. **Check the engine oil pressure.** (AM.III.G.S10)

16. **Identify and explain oil system flow on the assigned engine.** (AM.III.G.S10)

17. **Troubleshoot an engine oil pressure malfunction.** (AM.III.G.S11)

18. **Troubleshoot an engine oil temperature system.** (AM.III.G.S12)

19. **Identify types of metal found in an oil filter.** (AM.III.G.S13)

20. **Remove and inspect an engine chip detector.** (AM.III.G.S14)

H. Ignition and Starting Systems

References: AC 43.13-1; FAA-H-8083-32

Knowledge

1. **What is the main advantage of a magneto ignition system over a battery ignition system for an aircraft reciprocating engine?** (AM.III.H.K1)

   A magneto has its own source of electrical energy and is not dependent upon the battery.

2. **What is an all-weather spark plug?** (AM.III.H.K2)

   A shielded spark plug that has a recess in the shielding in which a resilient grommet on the ignition lead forms a watertight seal.

3. **What is meant by the reach of a spark plug?** (AM.III.H.K2)

   The length of the threads on the spark plug that screw into the cylinder head.

4. **What is the difference between a hot spark plug and a cold spark plug?** (AM.III.H.K2)

   A hot spark plug has a long path for the heat to travel between the nose core insulator and the spark plug shell. In a cold spark plug, the heat has a shorter distance to travel, and the spark plug operates cooler than a hot spark plug.

5. **What is the advantage of fine-wire spark plugs over massive electrode spark plugs?** (AM.III.H.K2)

   Fine-wire spark plugs have a firing end that is more open than that of a massive electrode spark plug. The open firing end allows the gases that contain lead to be purged from the spark plug so they will not form solid lead contaminates.

6. **Why is it important that the spark plugs be kept in numbered holes in a tray when they are removed from an engine?** (AM.III.H.K2)

   Spark plugs tell a good deal about the internal condition of the cylinders from which they were taken. By knowing the cylinder from which each spark plug came, the mechanic can take the proper action when a spark plug indicates such conditions as detonation or overheating.

7. **Why is it important that a torque wrench always be used when installing spark plugs in an aircraft engine?** (AM.III.H.K2)

   If the spark plugs are not put in tight enough, there is the possibility of a poor seal; if they are put in too tight, there is danger of cracking the insulation or damaging the threads on the spark plug or the cylinder.

8. **What kind of gauge should be used to measure the electrode gap in aircraft spark plugs?** (AM.III.H.K2)

   A round wire gauge.

9. **What is a shower of sparks ignition system?** (AM.III.H.K3)
   An ignition system that uses an induction vibrator to send pulsating DC into a set of retard breaker points on one of the magnetos. This provides a hot and retarded spark for starting the engine.

10. **What is an impulse coupling, and how does it work?** (AM.III.H.K3)
    An impulse coupling is a spring-loaded coupling between a magneto shaft and the drive gear inside the engine. When the engine is rotated for starting, the impulse coupling locks the magnet so it cannot turn. The spring in the coupling winds up as the crankshaft continues to turn, and when the piston is near top center, the coupling releases and spins the magnet, producing a hot and retarded spark.

11. **What are the three distinct circuits in a magneto system?** (AM.III.H.K4)
    The magnetic circuit, the primary electrical circuit, and the secondary electrical circuit.

12. **What are the advantages of a solid-state magneto?** (AM.III.H.K5)
    No mechanical parts that wear, higher energy spark at low speeds, and no recurring inspection requirements.

13. **What is the purpose of a FADEC system?** (AM.III.H.K6)
    FADEC (full-authority digital electronic control) is a digital electronic fuel control for a gas turbine engine that is functioning during all engine operations, hence full authority. It includes the EEC (electronic engine control) and functions with the flight management computer. FADEC schedules the fuel to the nozzles in such a way that prevents overshooting power changes and over-temperature conditions. FADEC furnishes information to the EICAS (engine indicating and crew alerting system).

14. **What kind of electric starting system is used on many of the smaller turbine engines?** (AM.III.H.K7)
    A starter-generator.

15. **Why are air-turbine starters superior to electric starters for large turbine engines?** (AM.III.H.K7)
    They are light weight for the torque they produce.

16. **Where does an air-turbine starter get its air for starting the engines on a jet transport airplane?** (AM.III.H.K7)
    From an APU, GPU, or from a running engine.

17. **What prevents too high an air pressure from overspeeding an air-turbine starter?** (AM.III.H.K7)
    The air shutoff and regulating valve.

18. **Where does an air-turbine starter get its lubricating oil?** (AM.III.H.K7)
    It has a self-contained lubrication system with the oil held in the starter housing.

19. **What device in an air-turbine starter warns an aviation mechanic if there are any metal chips or particles in the oil?** (AM.III.H.K7)
    Magnetic chip detectors warn of any metal contamination in the oil.

20. **What is the function of the capacitor in a magneto?** (AM.III.H.K8)

    The capacitor minimizes arcing at the breaker points, and it speeds up the collapse of the primary current as the breaker points open.

21. **What happens in a magneto ignition system when the ignition switch is placed in the Off position?** (AM.III.H.K8)

    The primary circuit is connected to ground.

22. **What is checked when a magneto is internally timed?** (AM.III.H.K8)

    Internally timing a magneto consists of adjusting the breaker points so they will open at the instant the rotating magnet is in its E-gap position, and the distributor rotor is in position to direct the high voltage to cylinder number one.

23. **In what position should the ignition switch be placed when using a timing light on the magnetos?** (AM.III.H.K8)

    In the Both position.

24. **What is the E-gap in magneto timing?** (AM.III.H.K8)

    The E-gap angle is the position of the rotating magnet where the primary current flowing in the magneto coil is the greatest. The breaker points open when the rotating magnet is in its E-gap position.

25. **What malfunction in the ignition system would cause an aircraft reciprocating engine to continue to run after the ignition switch is placed in the Off position?** (AM.III.H.K8)

    The ignition switch is not grounding the magneto primary circuit.

26. **What type of ignition system is used on most turbine engines?** (AM.III.H.K9)

    High-intensity, intermittent-duty, capacitor discharge ignition systems.

27. **How many igniters are used with most turbine engines?** (AM.III.H.K9)

    Two.

28. **What are two types of ignition systems used in turbine engines?** (AM.III.H.K9)

    High-voltage systems and low-voltage systems.

29. **With which type of turbine-engine ignition system is a glow plug igniter used?** (AM.III.H.K9)

    A low-voltage system.

30. **How is the strength of the magnet in a magneto checked?** (AM.III.H.K9)

    The magneto is put on a test stand and rotated at a specified speed. The breaker points are held open and the primary current is measured. The strength of the magnet determines the amount of primary current.

31. **In what position is the magnet in a magneto when the greatest change in flux density in the coil core takes place?** (AM.III.H.K9)

    It is a few degrees beyond its neutral position. When it is in this position, the breaker points open and the primary current is interrupted. The flux change in the coil core is the greatest.

32. **In what position is the magnet in a magneto when the breaker points begin to open?** (AM.III.H.K9)

    In its E-gap position, just a few degrees beyond its neutral position.

33. **What happens when the start switch for a turbine engine with a starter-generator is placed in the START position?** (AM.III.H.K9)

    a. Current flows into the starter relay which closes and actuates the ignition exciter.
    b. Current flows through the series motor coils of the starter generator and rotates the engine until it starts.

34. **What happens when the start switch for a turbine engine with a starter-generator is placed in the RUN position?** (AM.III.H.K9)

    a. The starter relay opens, shutting off current to the ignition exciter and the motor windings in the starter-generator.
    b. The generator field relay is closed, connecting the voltage regulator to the starter-generator.
    c. The output of the generator flows to the bus through the generator circuit breaker.

35. **What is indicated if the starter relay chatters rather than locks in when the start switch is placed in the START position?** (AM.III.H.K9)

    The battery voltage is too low to supply enough current to hold the relay engaged.

36. **What is the purpose of the shear section in an air-turbine drive shaft?** (AM.III.H.K9)

    If the ratchet mechanism fails to release and the engine drives the starter to a speed higher than its design speed, the shear section will break and disconnect the starter from the engine.

Risk Management

1. **What are the risks associated with incorrect retard points or incorrect impulse coupling (with a different lag angle) than what is specified by the manufacturer?** (AM.III.H.R1)

   Incorrectly retarded ignition can lead to a failure to start the engine, or severe backfiring of the engine, which can cause damage. Use the recommended timing procedures from the aircraft maintenance manual.

2. **What cautions should be taken when working around capacitor discharge ignition systems?** (AM.III.H.R2)

   Capacitor discharge systems are commonly used in turbine aircraft and produce a very high energy spark. If the capacitors in the system are still charged when the leads are disconnected, there is a danger of personnel encountering this high energy spark. Consult the maintenance manual for instructions on how long to wait after power has been disconnected before working on the system.

3. **What is the danger of working around reciprocating engines with an ungrounded magneto?** (AM.III.H.R3)

   Even if the propeller is turned very slowly, when the impulse coupling snaps, it sends a spark to the cylinder which can start the ignition process and start the engine. When working on an ungrounded magneto, stay clear of the propeller, never move the propeller, and, if possible, disconnect the ignition leads to the spark plugs.

Skills

1. **Demonstrate to the examiner how to set the internal timing of the supplied magneto.** (AM.III.H.S1)

2. **Internally time a magneto and install it on an engine specified by the examiner.** (AM.III.H.S2)

3. **Install and/or time a magneto on an engine.** (AM.III.H.S2)

4. **Check the timing of the magnetos to the engine and determine whether this timing is within the limits allowed by the engine manufacturer.** (AM.III.H.S2)

5. **Clean and gap a spark plug specified by the examiner. Test it and explain to the examiner the reason a spark plug will function properly in an engine if it sparks properly in the tester.** (AM.III.H.S3)

6. **Using the appropriate documentation, find the correct part number for the coil used in a magneto specified by the examiner.** (AM.III.H.S4)

7. **Demonstrate the correct way to inspect the ignition system of a turbine engine specified by the examiner.** (AM.III.H.S4)

8. **Demonstrate the way the brushes should be replaced in a starter-generator specified by the examiner.** (AM.III.H.S4)

9. **Disassemble, identify components, and reassemble a magneto.** (AM.III.H.S4)

10. **Fabricate an ignition lead.** (AM.III.H.S4)

11. **Test the high-tension leads.** (AM.III.H.S4)

12. **Check the P-lead in a reciprocating engine ignition system. Explain the reason you cannot kill the engine with the magneto switch if the P-lead is broken.** (AM.III.H.S5)

13. **Demonstrate to the examiner the correct way to inspect an electrical turbine engine starter for proper operation.** (AM.III.H.S5)

14. **Inspect an electric turbine-engine starter and explain to the examiner the way it should be lubricated.** (AM.III.H.S5)

15. **Check a magneto on a test bench.** (AM.III.H.S5)

16. **Check the serviceability of the condensers.** (AM.III.H.S5)

17. **Check the ignition coils.** (AM.III.H.S5)

18. **Check the ignition leads.** (AM.III.H.S5)

19. **Replace and adjust a set of breaker points in a magneto specified by the examiner.** (AM.III.H.S6)

20. **Inspect the magneto breaker points.** (AM.III.H.S6)

21. **What kind of equipment is used to test an ignition harness for a reciprocating engine?** (AM.III.H.S7)
    A high-voltage electrical leakage detector.

22. **Remove and install an ignition harness.** (AM.III.H.S7)

23. **Demonstrate to the examiner how to inspect an ignition harness.** (AM.III.H.S7)

24. **Inspect a magneto impulse coupling.** (AM.III.H.S8)

25. **Troubleshoot an electrical starting system.** (AM.III.H.S9)

26. **Troubleshoot an ignition switch circuit.** (AM.III.H.S10)

27. **Inspect and check gap of spark plugs.** (AM.III.H.S11)

28. **Identify the correct spark plugs used for replacement installation.** (AM.III.H.S12)

29. **Troubleshoot a turbine or reciprocating engine ignition system.** (AM.III.H.S13)

30. **Describe the correct way to remove the exciter box from a turbine engine specified by the examiner. Explain the safety procedures that must be observed.** (AM.III.H.S14)

31. **Identify the correct igniter plug and replace turbine engine igniter plugs.** (AM.III.H.S14)

32. **Troubleshoot a turbine engine's igniters.** (AM.III.H.S15)

33. **Inspect the turbine engine ignition system.** (AM.III.H.S16)

34. **Explain to the examiner the way a starter-generator starts a turbine engine and then shifts its action to that of a generator.** (AM.III.H.S16)

35. **Remove and inspect a low-voltage igniter plug from a turbine engine. Explain to the examiner the way the plug would be cleaned, tested, and replaced.** (AM.III.H.S17)

I. Engine Fuel and Fuel Metering Systems

References: AC 43.13-1; FAA-H-8083-32

Knowledge

1. **Why do aircraft fuel metering systems for reciprocating engines have mixture controls?** (AM.III.I.K1)

   The air density decreases as the aircraft ascends, causing the mixture to become richer. The mixture control allows the fuel to be decreased to maintain a fuel-air mixture ratio that produces the desired power.

2. **What does an air-fuel mixture ratio of 12:1 represent?** (AM.III.I.K1)

   This represents an air-to-fuel ratio mixture of 12 pounds of air and 1 pound of fuel.

3. **What is the function of a float carburetor?** (AM.III.I.K2)

   The float carburetor's function is to measure airflow through the engine induction system and dispense the appropriate amount of gasoline into the airflow for all engine operating perimeters and conditions. It must also provide the fuel in a state that is as vaporized as possible by the time ignition occurs in the engine cylinders.

4. **What two things are adjusted when setting the idling conditions on a float carburetor?** (AM.III.I.K2)

   a. Idling RPM by adjusting the throttle stop.
   b. Idling mixture by adjusting idle needle valve.

5. **What would likely cause a reciprocating engine equipped with a float carburetor to hesitate momentarily when the throttle is rapidly advanced from idle to full power?** (AM.III.I.K2)

   A malfunctioning accelerator pump.

6. **How does a pressure-type carburetor operate?** (AM.III.I.K3)

   A pressure-type carburetor discharges fuel into the airstream at a pressure well above atmospheric. This results in better vaporization and permits the discharge of fuel into the airstream on the engine side of the throttle valve. With the discharge nozzle located at this point, the drop in temperature due to fuel vaporization takes place after the air has passed the throttle valve and at a point where engine heat tends to offset it. Thus, the danger of fuel vaporization icing is practically eliminated. The effects of rapid maneuvers and rough air on the pressure-type carburetors are negligible since its fuel chambers remain filled under all operating conditions.

7. **What are the two types of fuel injection systems used on modern reciprocating aircraft engines?** (AM.III.I.K4)

   Teledyne Continental (TCM) system and Precision Airmotive (Bendix) system.

8. **What is used in the TCM fuel injection system to control the amount of fuel sent to the nozzles?** (AM.III.I.K4)

   The engine speed determines the injector pump output pressure. The relief valve determines the fuel for low-speed operation, and the adjustable orifice determines the fuel for high-speed operation.

9. **What is used in the Precision Airmotive system to control the amount of fuel sent to the nozzles?** (AM.III.I.K4)

   The mass of the air entering the engine acts on an air diaphragm that controls a servo ball valve in the line between the regulator and the flow divider.

10. **What is a FADEC?** (AM.III.I.K5)

    A full-authority digital electronic control (FADEC) is a high-precision digital electronic fuel control that functions during all engine operations. It includes the electronic engine control (EEC) and functions with the flight management computer (FMC) to measure the fuel to the nozzles in such a way that prevents overshooting of power changes and over-temperature conditions. FADEC furnishes information to the engine indicating and crew alerting system (EICAS).

11. **Does a turbine engine controlled by a FADEC require manual trimming?** (AM.III.I.K5)

    No, this is done by the FADEC.

12. **What are the basic principles of a hydromechanical fuel control?** (AM.III.I.K6)

    A pure hydromechanical fuel control has no electronic interface assisting in computing or metering the fuel flow. It also is generally driven by the gas generator gear train of the engine to sense engine speed. Other mechanical engine parameters that are sensed are compressor discharge pressure, burner pressure, exhaust temperature, and inlet air temperature and pressure. Once the computing section determines the correct amount of fuel flow, the metering section through cams and servo valves delivers the fuel to the engine fuel system.

13. **What are the two basic types of fuel nozzles?** (AM.III.I.K7)

    Simplex and duplex configurations.

14. **Why does a duplex fuel nozzle usually require a dual fuel manifold?** (AM.III.I.K7)

    The duplex nozzle usually requires a dual manifold and a pressurizing valve or flow divider for dividing primary and secondary (main) fuel flow.

15. **What is the advantage of a duplex fuel nozzle?** (AM.III.I.K7)

    The duplex nozzle offers a desirable spray pattern for combustion over a wide range of operating pressures.

16. **What parameters are normally sensed by a turbine-engine fuel control?** (AM.III.I.K8)

    a. Power lever angle.
    b. Compressor inlet total temperature.
    c. Compressor RPM.
    d. Burner pressure.

17. What is meant by trimming a turbine engine? (AM.III.I.K8)

Adjusting the fuel control for the proper specific gravity of the fuel and for the proper idle and full throttle RPM.

18. What are two types of electronic fuel controls for turbine engines? (AM.III.I.K8)

Supervisory electronic engine control and full-authority digital electronic control.

19. What position should the flight deck fuel metering control (or fuel condition lever) be set to for starting a turbine engine? (AM.III.I.K8)

The normal starting sequence is:

1. Rotate the compressor with the starter;
2. Turn the ignition on; and
3. Open the engine fuel valve, either by moving the throttle to idle or by moving a fuel shutoff lever or turning a switch.

20. What specifications are used to trim a turbine engine? (AM.III.I.K9)

The data plate specifications for the specific engine. This data was obtained when the engine was calibrated in the manufacturer's test cell.

21. Where should ambient temperature be measured when trimming a turbine engine? (AM.III.I.K9)

In a shaded area as near the engine inlet as is practical.

22. When should a turbine engine be retrimmed? (AM.III.I.K9)

Any time there is a decrease in engine thrust, and after any maintenance that the manufacturer specifies as requiring retrimming.

23. Are the instruments installed in an aircraft suitable for use when trimming a turbine engine? (AM.III.I.K10)

No, a special analyzer/trimmer such as a JetCal should be used.

24. Where are the fuel strainers normally located in an aircraft fuel system? (AM.III.I.K10)

Fuel strainers are located in the tank outlet, the main strainer is located at the lowest point in the fuel system, and there are strainers in the carburetor or fuel injection system.

25. How can you tell when a fuel strainer for a turbine engine has been bypassed? (AM.III.I.K10)

There is an indicator button that will pop out. This button is usually painted red for easy identification of a bypass condition.

26. What are two types of fuel heaters used in a jet transport aircraft? (AM.III.I.K11)

Air-to-fuel and oil-to-fuel heat exchangers.

27. How are fuel lines often protected in an engine compartment? (AM.III.I.K12)

By wrapping the lines with a fire protection sleeve to protect from fire and abrasion.

28. What is the purpose of a lay line (solid line or printed words) on a flexible fuel hose? (AM.III.I.K12)

To monitor fuel hose twist. Hoses should be installed without twisting.

29. **What kind of fuel boost pumps are normally installed inside the fuel tanks?** (AM.III.I.K13)

    Centrifugal pumps.

30. **What are three functions of the boost pump in the tank of a jet transport aircraft?** (AM.III.I.K13)

    a. To pressurize the fuel in the line between the tank and the engine-driven pump.
    b. To transfer fuel from one tank to another to balance the fuel load.
    c. To pump fuel from the tank into the dump chute when fuel is being dumped.

31. **What is meant by a compensated fuel pump?** (AM.III.I.K13)

    It is a fuel pump that senses the ambient air pressure and maintains the fuel pressure a specific amount above this air pressure.

32. **What are the two types or categories of engine driven fuel pumps?** (AM.III.I.K13)

    Nonconstant displacement and nonpositive displacement.

33. **What are the three basic types of valves used in fuel systems?** (AM.III.I.K14)

    Cone-type, poppet-type, and gate valves.

34. **What are the three most common types of fuel filters?** (AM.III.I.K15)

    The micron filter, the wafer screen filter, and the plain screen mesh filter.

35. **Why do some aircraft filters have a relief valve?** (AM.III.I.K15)

    If the fuel screen, or filter, becomes clogged, the relief valve will allow fuel to continue to flow to the engine.

36. **Where are fuel drains located in an aircraft engine?** (AM.III.I.K16)

    On a turbine engine, combustion chambers have a drain valve that drains fuel that accumulates in the combustion chamber after each shutdown, or any fuel that may have accumulated during a false start. Many reciprocating engines have drain valves in the intake manifold that open when the engine is not running. These drains remove excess fuel if the engine is flooded from over-priming or an unsuccessful engine start.

Risk Management

1. **What risks are associated with incorrect adjustments of a turbine engine fuel control unit (FCU)?** (AM.III.I.R1)

   Incorrect adjustments of a fuel control unit can lead to poor engine performance and the inability to obtain full takeoff power. The aircraft maintenance manual procedures must be followed completely to ensure correct rigging of the FCU.

2. **What problems can occur if the air-fuel control unit on a reciprocating engine is not adjusted correctly?** (AM.III.I.R2)

   If the fuel control unit is not adjusted correctly, the engine may not reach full power, the engine could run rough, or the engine could die at low power settings due to incorrect air-fuel ratios.

3. **What dangers are present when the handling of fuel metering system components results in fuel spillage on the hangar floor?** (AM.III.I.R3)

   Fuel spilled on the floor represents a danger from the fuel fumes, is a hazard for personnel to slip on, and has a deteriorating effect on tires that are spilled on or that roll through fuel. Any tire that is contaminated should be cleaned with soap and water and dried. Fuel spills should be addressed immediately.

4. **What considerations or precautions should be taken when performing maintenance on fuel systems?** (AM.III.I.R4)

   Care should be taken to not breathe in fuel vapors and not allow fuel to remain on exposed skin. Use proper PPE when working on fuel systems. Spilled fuels should be cleaned up immediately. Fuel should not be allowed to drip on paint, tires, or any rubber components as damage can occur.

5. **What risks are associated with spilling fuel when handling a fuel control unit?** (AM.III.I.R5)

   Fuel spilled on the floor represents a danger from the fuel fumes, is a hazard for personnel to slip on, and has a deteriorating effect on tires that are spilled on or that roll through fuel. Any tire that is contaminated should be cleaned with soap and water and dried. Fuel spills should be addressed immediately.

Skills

1. **Inspect, troubleshoot, and repair a continuous-flow fuel injection system.** (AM.III.I.S1)

2. **Replace a direct-injection fuel nozzle.** (AM.III.I.S2)

3. **Remove, inspect, and install a turbine engine fuel nozzle.** (AM.III.I.S2)

4. **Identify the components of a carburetor.** (AM.III.I.S3)

5. **Identify the main discharge nozzle in a pressure carburetor.** (AM.III.I.S3)

6. **Interpret a diagram showing fuel and air flow through a float-type and/or a pressure-type carburetor.** (AM.III.I.S4)

7. **Identify a carburetor's air-bleed system.** (AM.III.I.S4)

8. **Remove and/or install a main metering jet in a carburetor.** (AM.III.I.S5)

9. **Service a carburetor fuel inlet screen.** (AM.III.I.S6)

10. **Adjust a continuous-flow fuel injection system.** (AM.III.I.S7)

11. **Inspect a float needle and/or seat in a float-type carburetor.** (AM.III.I.S8)

12. **Check the float level on a float-type carburetor.** (AM.III.I.S8)

13. **Identify, remove, and/or install a float-type carburetor.** (AM.III.I.S9)

14. **Remove and/or install the mixture control system in a float-type carburetor.** (AM.III.I.S9)

15. **Remove and/or install the accelerating pump in a float-type carburetor.** (AM.III.I.S9)

16. **Adjust the idle speed and/or mixture.** (AM.III.I.S10)

17. **Demonstrate the correct way to adjust the idle speed and mixture on an engine specified by the examiner.** (AM.III.I.S10)

18. **Describe the conditions that may result in turbine engine RPM overspeed.** (AM.III.I.S11)

19. **Locate procedures for a turbine engine revolutions per minute (RPM) overspeed inspection.** (AM.III.I.S11)

20. **Inspect fuel metering flight deck controls for proper adjustment.** (AM.III.I.S12)

21. **Locate procedures for adjusting a hydromechanical fuel control unit.** (AM.III.I.S13)

22. **Inspect a turbine fuel control unit.** (AM.III.I.S14)

23. **Locate and explain procedures for removing and installing a turbine engine fuel control unit.** (AM.III.I.S14)

24. **Identify components of an engine fuel system.** (AM.III.I.S15)

25. **Remove and/or install an engine-driven fuel pump.** (AM.III.I.S16)

26. **Rig a remotely operated fuel valve.** (AM.III.I.S17)

27. **Explain to the examiner the way to check a remotely located fuel selector valve for proper operation and freedom of action.** (AM.III.I.S17)

28. **Locate and identify the fuel selector placards.** (AM.III.I.S18)

29. **Demonstrate the correct way to remove and clean a main fuel filter and check it for leaks after it has been reinstalled.** (AM.III.I.S19)

30. **Demonstrate the correct way to check a fuel boost pump to determine that it is putting out the pressure specified in the aircraft maintenance manual.** (AM.III.I.S20)

31. **Remove and/or install a fuel boost pump.** (AM.III.I.S20)

32. **Locate and identify a turbine engine fuel heater.** (AM.III.I.S21)

33. **Check the fuel pressure warning light function.** (AM.III.I.S22)

34. **Adjust fuel pumps fuel pressure.** (AM.III.I.S23)

35. **Inspect the engine fuel system fluid lines and/or components.** (AM.III.I.S24)

36. **Demonstrate to the examiner the way to check an engine fuel system for leaks using pressure from the auxiliary fuel pump.** (AM.III.I.S24)

37. **Troubleshoot abnormal fuel pressure.** (AM.III.I.S25)

38. **Troubleshoot a turbine engine fuel heater system.** (AM.III.I.S26)

39. **Locate the procedures for troubleshooting a turbine engine fuel heater system.** (AM.III.I.S26)

40. **Remove, clean, and/or replace an engine fuel strainer.** (AM.III.I.S27)

41. **Troubleshoot engine fuel pressure fluctuation.** (AM.III.I.S28)

42. **Inspect the fuel selector valve.** (AM.III.I.S29)

43. **Demonstrate the way to inspect the fuel nozzle on a turbine engine for the correct spray pattern.** (AM.III.I.S30)

J. Reciprocating Engine Induction and Cooling Systems

References: AC 43.13-1; FAA-H-8083-32

Knowledge

1. **What does a fuel-injected engine have that prevents the loss of induction air if the air inlet filter should become covered with ice?** (AM.III.J.K1)

   An alternate air valve allows warm air from inside the engine cowling to flow into the fuel metering system.

2. **What would be the effect on engine operation of an air leak in the intake pipe for one cylinder?** (AM.III.J.K1)

   That cylinder would run lean and detonation could occur.

3. **What must be inspected on the cooling system of an air-cooled engine?** (AM.III.J.K1)

   All of the baffles and air seals must be in place and in good condition.

4. **Why should an air-cooled engine not be run up to high power without the cowling installed?** (AM.III.J.K1)

   The engine depends on the cowling to force air through the cylinder fins to remove the excess heat.

5. **Why do some air-cooled engine installations have cowl flaps?** (AM.III.J.K1)
   Cowl flaps produce a low pressure on one side of the engine to pull air through the cooling fins on the cylinders.

6. **What document describes the proper use of cowl flaps on an air-cooled engine?** (AM.III.J.K1)
   The pilot's operating handbook.

7. **Why does ice form in the throat of a float carburetor?** (AM.III.J.K2)
   When liquid fuel evaporates it absorbs enough heat from the air to cause moisture to condense out and freeze.

8. **Are internally driven superchargers located before or after the carburetor?** (AM.III.J.K3)
   Internally driven superchargers compress the air-fuel mixture after it leaves the carburetor, while externally driven superchargers (turbochargers) compress the air before it is mixed with the metered fuel from the carburetor.

9. **What are the three main parts of a typical turbosupercharger?** (AM.III.J.K4)
   Compressor assembly, turbine wheel assembly, and full floating shaft bearing assembly.

10. **How is a turbosupercharger (or turbocharger) driven?** (AM.III.J.K4)
    Externally driven superchargers derive their power from the energy of engine exhaust gases directed against a turbine that drives an impeller that compresses the incoming air.

11. **How is a supercharger driven?** (AM.III.J.K4)
    The supercharger is directly driven from the engine and is internal to the engine.

12. **How does an augmentor system increase cooling airflow over a reciprocating engine?** (AM.III.J.K5)
    Augmentors use exhaust gas velocity to cause airflow over the engine so that cooling is not entirely dependent on the prop wash.

13. **Why is it very important that the induction air filters be kept clean and replaced as often as the manufacturer recommends?** (AM.III.J.K6)
    Clogged air filters can restrict the air entering the engine.

14. **Does the application of carburetor heat cause the fuel-air mixture to become richer or leaner?** (AM.III.J.K7)
    The less dense, heated air draws the same amount of fuel from the carburetor as cold air; therefore the mixture becomes richer.

15. **What happens to engine RPM when carburetor heat is applied?** (AM.III.J.K7)
    The RPM drops.

16. **Why should the use of carburetor heat be limited when operating an engine with the aircraft on the ground?** (AM.III.J.K7)
    The air that flows into the engine when carburetor heat is applied is not filtered.

17. **Why is a fuel-injected reciprocating engine not as prone to icing as an engine equipped with a float carburetor?** (AM.III.J.K7)
    In a fuel-injected engine, the liquid fuel evaporates in the intake valve chamber of the hot cylinder head. In a carbureted engine, as air flows through the carburetor venturi, the pressure drops, decreasing temperature and increasing the chance of icing.

18. **What creates the pressure of the air in a pressure cowling?** (AM.III.J.K8)
    The forward velocity of the airplane as air enters openings in the cowling and flows into the engine compartment.

19. **What is meant by pressure cooling of an air-cooled engine?** (AM.III.J.K8)
    Ram air on one side of the engine is forced by a series of baffles to flow through the fins on the cylinders to a low-pressure area on the other side of the engine.

20. **What controls the amount of air cooling in a pressure cowling system?** (AM.III.J.K8)
    The exit air is controlled through the use of cowl flaps. The cowl flaps are opened to allow more airflow and more cooling. The cowl flaps are closed to reduce the amount of airflow over the engine.

21. **What is the purpose of engine baffles and seals in a pressure cowling system?** (AM.III.J.K9)
    The baffles and seals cause the air to be directed over the cooling fins on the cylinders.

22. **Where can you find what is the maximum amount of cylinder fin area that can be removed in order to clean out a damaged area?** (AM.III.J.K9)
    In the engine overhaul manual.

23. **How is heat removed from the liquid in a liquid-cooled engine?** (AM.III.J.K10)
    By flowing air over a radiator.

Risk Management

1. **Why is it important to follow proper maintenance procedures on a turbocharger?** (AM.III.J.R1)
   Turbochargers rotate at high speeds and generate significant amounts of heat. Proper oil flow and control through a turbocharger is critical for cooling and control of the turbocharger system. Improper maintenance leads to improper operation and early failure of the system.

2. **What are some of the risks associated with ground operation of aircraft engines?** (AM.III.J.R2)
   The cooling of aircraft engines is optimized for inflight operations. Ground operations do not provide sufficient air to provide the same amount of cooling, and engines can overheat. Follow maintenance manual procedures for ground operations and monitor engine temperatures to prevent the engine from overheating.

3. **What are the risks of maintenance-related FOD, and how can it be prevented?** (AM.III.J.R3)

   Maintenance-related FOD can come in many different forms. Tools that are laid down on top of an engine or in the inlet duct of a turbine engine can quickly enter the engine and cause major damage to the engine. Other sources of FOD include hardware, safety wire, protective caps, and items falling out of aviation mechanic's pockets, all of which can cause damage to an engine.

4. **What cautions are to be taken when servicing a liquid cooling system?** (AM.III.J.R4)

   The chemicals used in coolants can be harmful if ingested and can cause irritation to exposed skin. Use proper PPE when working on cooling systems. Coolant can also be harmful to tires and rubber components on the engine. Clean up spilled coolant immediately.

Skills

1. **Demonstrate the correct way to check for proper functioning of a carburetor heat system on an operating engine.** (AM.III.J.S1)

2. **Troubleshoot an engine that idles poorly.** (AM.III.J.S1)

3. **Troubleshoot an engine that fails to start.** (AM.III.J.S1)

4. **Troubleshoot a carburetor heat system.** (AM.III.J.S1)

5. **Explain to the examiner the probable locations of induction ice.** (AM.III.J.S1)

6. **Demonstrate the correct way to inspect a carburetor heat system for proper movement of the air valve, and proper cushion on the control.** (AM.III.J.S1)

7. **Inspect an alternate air valve for proper operation.** (AM.III.J.S2)

8. **Inspect an induction system drain for proper operation.** (AM.III.J.S3)

9. **Inspect an engine exhaust augmenter cooling system.** (AM.III.J.S4)

10. **Demonstrate the correct way to clean an induction air filter specified by the examiner.** (AM.III.J.S5)

11. **Check a turbocharger for operation.** (AM.III.J.S6)

12. **Inspect a turbocharger for leaks and security.** (AM.III.J.S6)

13. **Inspect and service a turbocharger waste gate.** (AM.III.J.S7)

14. **Inspect an induction system for obstruction.** (AM.III.J.S8)

15. **Demonstrate the correct way to check an induction system for air leaks.** (AM.III.J.S9)

16. **Demonstrate how to correct a leak in an intake pipe to a cylinder.** (AM.III.J.S9)

17. **Locate the proper specifications for coolant used in a liquid-cooled engine.** (AM.III.J.S10)

18. **Demonstrate the correct way to inspect the cooling system of an engine specified by the examiner.** (AM.III.J.S11)

19. **Explain to the examiner the method of heat transfer used to remove heat from an engine cylinder.** (AM.III.J.S11)

20. **Explain to the examiner the way a missing inter-cylinder baffle can cause damage to an engine cylinder.** (AM.III.J.S11)

21. **Identify the components in the intake side of a turbocharged engine specified by the examiner.** (AM.III.J.S12)

22. **Identify components of a turbocharger induction system.** (AM.III.J.S12)

23. **Identify exhaust augmenter-cooled engine components.** (AM.III.J.S13)

24. **Inspect an air inlet duct for security.** (AM.III.J.S14)

25. **Perform an induction and cooling system inspection.** (AM.III.J.S15)

26. **Repair a cylinder baffle.** (AM.III.J.S16)

27. **Explain to the examiner the function of the baffles and seals in an engine compartment.** (AM.III.J.S17)

28. **Inspect cylinder baffling.** (AM.III.J.S17)

29. **Demonstrate the correct way to inspect a cowl flap installation and explain the way to adjust it.** (AM.III.J.S18)

30. **Explain to the examiner the reason engine cowl flaps are normally required to be open when operating on the ground.** (AM.III.J.S18)

31. **Repair a cylinder cooling fin.** (AM.III.J.S19)

32. **Inspect cylinder cooling fins.** (AM.III.J.S19)

K. Turbine Engine Air Systems

References: AC 43.13-1; FAA-H-8083-32

Knowledge

1. **How is a turbine engine cooled?** (AM.III.K.K1)
   A turbine engine is cooled by the air passing over and through the engine. On a turbofan engine, a large portion of the air flows over the outside of the engine, helping to cool the various sections of the engine.

2. **Why is it important for turbine engines to have a relatively distortion-free flow of air to the inlet of the compressor?** (AM.III.K.K2)
   A uniform and steady airflow is necessary to avoid compressor stall (airflow tends to stop or reverse direction of flow) and excessive internal engine temperatures in the turbine section.

3. **How are the high internal temperatures of a turbine engine cooled and kept under control?** (AM.III.K.K3)
   A portion of the air flowing into the engine flows around the combustion chamber, and even around the internal combustion flame, to contain the flame and protect internal components of the engine.

4. **What are baffles used for in the air flow of a turbine engine?** (AM.III.K.K4)
   To direct air-cooling air to protect engine and airframe components. The combustion chamber is a specialized baffle that directs compressed air into the combustion chamber and helps contain the flame.

5. **What is the purpose of insulation blankets in a turbine engine installation?** (AM.III.K.K5)
   Although insulation blankets protect the fuselage from heat radiation, they are used primarily to reduce heat loss from the exhaust system. The reduction of heat loss improves engine performance.

6. **What is the primary design concern of a turbine engine inlet?** (AM.III.K.K6)
   The air entrance is designed to conduct incoming air to the compressor with a minimum energy loss resulting from drag or ram pressure loss; that is, the flow of air into the compressor should be free of turbulence to achieve maximum operating efficiency. Proper inlet design contributes materially to aircraft performance by increasing the ratio of compressor discharge pressure to duct inlet pressure.

7. **What is the compressor pressure ratio?** (AM.III.K.K6)
   This ratio is the outlet pressure divided by the inlet pressure. The amount of air passing through the engine is dependent upon three factors:
   1. The compressor speed (RPM).
   2. The forward speed of the aircraft.
   3. The density of the ambient (surrounding) air.

8. **Where is bleed air taken from on a turbine engine?** (AM.III.K.K7)

   Bleed air is taken from any of the various pressure stages of the compressor. Air is often bled from the final or highest pressure stage since, at this point, pressure and air temperature are at a maximum.

9. **What are some uses of bleed air?** (AM.III.K.K7)

   a. Cabin pressurization, heating, and cooling
   b. Deicing and anti-icing equipment
   c. Pneumatic starting of engines
   d. Auxiliary drive units (ADU)
   e. Vortex dissipaters

10. **How is ice prevented from forming on the nose cowl, nose dome, and inlet guide vanes of a turbine engine?** (AM.III.K.K8)

    Hot compressor bleed air flows through passages in these components to elevate the surface temperatures sufficiently to prevent ice from forming.

Risk Management

1. **Why is it important to follow manufacturer's instructions when maintaining compressor bleed air systems?** (AM.III.K.R1)

   Poor maintenance can lead to bleed-air valve malfunctions, which can cause high exhaust gas temperatures during takeoff, climb, and cruise with higher-than-normal RPMs and fuel flows.

2. **What risks are associated with ground operations following other than the manufacturer's instructions?** (AM.III.K.R2)

   Failure to follow the manufacturer's instructions can lead to safety issues and hot or hung starts. The manufacturer's operating instructions should be consulted before attempting to start and operate any turbine engine. Prior to start, particular attention should be paid to the engine air inlet, the visual condition and free movement of the compressor and turbine assembly, and the parking ramp area fore and aft of the aircraft. During the start, it is necessary to monitor the tachometer, the oil pressure, and the exhaust gas temperature. The normal starting sequence is:
   1. Rotate the compressor with the starter;
   2. Turn the ignition on; and
   3. Open the engine fuel valve, either by moving the throttle to idle or by moving a fuel shutoff lever or turning a switch.

Skills

1. **Perform an induction and cooling system inspection.** (AM.III.K.S1)

2. **Troubleshoot an engine cooling system.** (AM.III.K.S1)

3. **Identify the exhaust augmentor cooled engine components.** (AM.III.K.S1)

4. **Troubleshoot a rotorcraft engine's cooling system.** (AM.III.K.S1)

5. **Identify the rotorcraft engine cooling components.** (AM.III.K.S1)

6. **Inspect a rotorcraft engine's cooling system.** (AM.III.K.S1)

7. **Inspect an engine exhaust augmentor cooling system.** (AM.III.K.S1)

8. **Identify the location of turbine engine insulation blankets.** (AM.III.K.S2)

9. **Repair the turbine engine insulation blankets.** (AM.III.K.S2)

10. **Identify turbine engine cooling air flow.** (AM.III.K.S3)

11. **Identify the turbine engine cooling air flow.** (AM.III.K.S4)

12. **Inspect turbine engine cooling ducting (rigid or flexible) or baffle seals.** (AM.III.K.S4)

13. **Inspect a turbine engine air intake anti-ice system.** (AM.III.K.S5)

14. **Identify the components in the anti-icing system in a turbine engine installation specified by the examiner.** (AM.III.K.S6)

15. **Identify to the examiner the turbine engine air intake ice-protected areas.** (AM.III.K.S6)

16. **Inspect a particle separator.** (AM.III.K.S7)

17. **Inspect/check a bleed air system.** (AM.III.K.S8)

L. Engine Exhaust and Reverser Systems

References: AC 43.13-1; FAA-H-8083-32

Knowledge

1. **How does a leak appear in an engine exhaust system?** (AM.III.L.K1)
   It normally looks like a gray or black feather-shaped streak coming from a crack or from a location where components are not in perfect alignment.

2. **How can you check a reciprocating engine exhaust system for leaks?** (AM.III.L.K1)
   Pressurize the exhaust system with the discharge from a vacuum cleaner and wipe a soap solution over all joints and suspect areas. A leak will cause bubbles to form.

3. **How do the components in the exhaust system of a turbocharged engine allow for expansion and contraction due to heat, and at the same time prevent leakage?** (AM.III.L.K1)

   Bellows and ball joints prevent leakage while allowing contraction and expansion.

4. **Should a turbocharged engine be started with the waste gate open or closed?** (AM.III.L.K1)

   Open, so as much exhaust as possible will bypass the turbine.

5. **What is the primary function of a reciprocating engine exhaust system?** (AM.III.L.K1)

   Its main function is to dispose of the gases with complete safety to the airframe and the occupants of the aircraft. The exhaust system can perform many useful functions, but its first duty is to provide protection against the potentially destructive action of the exhaust gases.

6. **What is a short stack exhaust system?** (AM.III.L.K1)

   The short stack exhaust system utilizes open exhaust tubes with no mufflers. The short stack system is generally used on non-supercharged engines and low-powered engines where the noise level is not too objectionable.

7. **How is exhaust flow different in a ducted fan turbine engine verses an unducted fan engine?** (AM.III.L.K2)

   Ducted fan engines take the fan airflow and direct it through closed ducts along the engine. Then, it flows into a common exhaust nozzle. The core exhaust flow and the fan flow mix and flow from the engine through this mixed nozzle. The unducted fan has two nozzles, one for the fan airflow and one for the core airflow. These both flow to ambient air separate from each other and have separate nozzles.

8. **What is the purpose of an augmentor on a reciprocating engine exhaust system?** (AM.III.L.K3)

   The augmentors are designed to produce a venturi effect to draw an increased airflow over the engine to augment engine cooling.

9. **How does a reciprocating engine muffler decrease engine noise?** (AM.III.L.K3)

   Exhaust gases from the cylinders are combined into a common manifold that contains a section of the exhaust where the volume increases (expansion chamber), slowing the gases down, diffusing the flow through internal baffles, and reducing the noise.

10. **What is a jet aircraft hush kit?** (AM.III.L.K3)

    A hush kit is an aerodynamic device that defuses the flow of exhaust from older turbojet engines, by reducing the noise caused by high-velocity jet exhaust.

11. **Why are thrust reversers not normally used when the airplane is moving below approximately 60 knots?** (AM.III.L.K4)

    There is a danger of recirculating the exhaust gases and of the engine ingesting foreign objects stirred up by the high-velocity gases.

12. **What is a mechanical-blockage thrust reverser?** (AM.III.L.K4)
    A thrust reverser that slides a pair of scoop-shaped doors aft and opens them so they block the rearward flow of gases and deflect the gases forward.

13. **How does a cascade-type thrust reverser operate?** (AM.III.L.K4)
    A portion of the fan cowl moves rearward, and a series of blocker doors deflect the fan discharge air through fan cascades that direct the discharge air forward. At the same time, blocker doors shut off the flow of gases from the core engine and deflect it through cascades that direct it forward.

Risk Management

1. **What risks are associated with exhaust system maintenance and inspection on reciprocating engines?** (AM.III.L.R1)
   Exhaust systems become very hot during operation, and the components will remain heat-soaked after shutdown. Wait for components to cool down sufficiently before working on them and wear proper gloves to protect against burns. Another challenge of inspecting exhaust systems is that internal baffles and diffusers can be difficult or impossible to see without partial disassembly of the exhaust system. It is important to understand the exhaust system and understand the procedures needed for proper inspection.

2. **What dangers are associated with performing maintenance on thrust reversers?** (AM.III.L.R2)
   Thrust reversers can move quickly with much force. Personnel must remain clear during ground checks, and measures should be put in place to ensure that the system is not inadvertently energized.

3. **What dangers exist with operating reciprocating engines with exhaust leaks?** (AM.III.L.R3)
   Exhaust leaks are particularly dangerous, as hot gases can cause damage to aircraft structure and other engine systems. If the exhaust leak is in a cabin heat exchange area, carbon monoxide can enter the cabin and render the occupants unconscious.

4. **What type of damage is most common in the exhaust system of a reciprocating engine?** (AM.III.L.R4)
   Cracks caused by expansion and contraction.

5. **What is the danger of an internal failure of baffles and diffusers in an exhaust muffler?** (AM.III.L.R4)
   Internal failures can cause sections of baffles and diffusers to break off and block the exit of the exhaust, causing loss of engine power. During inspection of exhaust systems, it is important to inspect the internal portions of the mufflers.

6. **What risks are associated with exhaust systems during ground operation of aircraft engines?** (AM.III.L.R5)

   Aircraft engines are designed for operation in flight. Ground operations do not provide the same amount of cooling of exhaust systems as when the aircraft is in flight. Extended ground operations can introduce overheat conditions that are difficult to detect and may cause damage to the engine and exhaust systems. Follow maintenance manual procedures for ground run operations to reduce risk of damage.

Skills

1. **Identify for the examiner the components in a turbocharger system, and explain the way the system operates.** (AM.III.L.S1)

2. **Identify the type of exhaust system on a particular aircraft.** (AM.III.L.S1)

3. **Clean the exhaust system components.** (AM.III.L.S1)

4. **Inspect the wastegate and the controller for a turbocharged engine and explain the way the controller changes the wastegate position.** (AM.III.L.S1)

5. **Inspect a turbine engine exhaust nozzle.** (AM.III.L.S2)

6. **Inspect a turbine engine exhaust system component provided by the examiner.** (AM.III.L.S2)

7. **Demonstrate the correct way to inspect a reciprocating engine exhaust system for leaks.** (AM.III.L.S3)

8. **Demonstrate the proper way to replace exhaust gaskets on an engine specified by the examiner. Explain the precautions that must be taken when working on the exhaust system.** (AM.III.L.S3)

9. **Remove and install the exhaust ducts.** (AM.III.L.S3)

10. **Repair an exhaust system leak.** (AM.III.L.S3)

11. **Demonstrate the way to mark an exhaust component for repair. Explain why lead pencils should not be used for marking.** (AM.III.L.S3)

12. **Inspect an exhaust system's internal baffles or diffusers.** (AM.III.L.S4)

13. **Inspect an exhaust heat exchanger.** (AM.III.L.S5)

14. **Remove and install a heat exchanger collector tube.** (AM.III.L.S5)

15. **Troubleshoot an exhaust muffler heat exchanger.** (AM.III.L.S5)

16. **Explain to the examiner the way to check the thrust reverser for security of mounting and for cracks or other forms of heat damage.** (AM.III.L.S6)

17. **Explain to the examiner the repairs that can be made to a thrust reverser.** (AM.III.L.S6)

18. **Troubleshoot a thrust reverser system.** (AM.III.L.S6)

19. **Demonstrate to the examiner the way a thrust reverser operates.** (AM.III.L.S6)

20. **Perform a heat exchanger collector tube leak test.** (AM.III.L.S7)

21. **Perform a pressure leak check of a reciprocating engine exhaust system.** (AM.III.L.S7)

M. Propellers

References: AC 43.13-1; FAA-H-8083-32

Knowledge

1. **Why do some aircraft engines have a critical range of operation?** (AM.III.M.K1)
   The engine and propeller combination have a resonant frequency problem in which excessive vibration can occur in a certain range of RPM. The tachometer for an aircraft with this limitation is marked with a red arc. When it is necessary to pass through this range it must be done as quickly as practical.

2. **Does centrifugal twisting moment on a propeller blade tend to move the blades toward high pitch or toward low pitch?** (AM.III.M.K1)
   Toward low pitch.

3. **Do the counterweights on a propeller tend to move the blades toward high pitch or toward low pitch?** (AM.III.M.K1)
   Toward high pitch.

4. **What is the function of the accumulator used with some McCauley feathering propellers?** (AM.III.M.K1)
   The accumulator stores oil under pressure when the engine is operating normally. This oil is used to help the propeller blades move toward low pitch when the propeller is being unfeathered.

5. **What is meant by the alpha range of operation of a turboprop propeller?** (AM.III.M.K1)
   This is the in-flight mode of operation from takeoff to landing.

6. **Should an adjustable-pitch propeller be in high pitch or in low pitch for takeoff?** (AM.III.M.K1)
   In low pitch.

7. **What can be done to prevent the front cone from bottoming when installing a propeller on a splined shaft?** (AM.III.M.K1)

   Install a spacer behind the rear cone to move the propeller forward on the shaft.

8. **Why is a propeller indexed to the engine crankshaft?** (AM.III.M.K1)

   The relationship between the propeller and the engine crankshaft is chosen to produce the minimum vibration.

9. **What is the function of the snap ring inside the hub of a propeller that is mounted on a tapered or splined shaft?** (AM.III.M.K2)

   The snap ring allows the propeller to be pulled off of the shaft when the retaining nut is backed off.

10. **What are the basic types of propellers?** (AM.III.M.K2)

    Fixed pitch, ground adjustable, constant-speed non-feathering, and constant-speed feathering.

11. **What materials are propeller blades made from?** (AM.III.M.K2)

    Wood, aluminum, and composite construction.

12. **What is adjusted inside the governor for a constant-speed propeller to change the speed at which the propeller is operating?** (AM.III.M.K3)

    The compression of the speeder spring.

13. **What is done to cause a McCauley propeller to feather?** (AM.III.M.K3)

    Oil is allowed to drain out of the propeller.

14. **What keeps a McCauley feathering propeller from feathering when the engine is shut down on the ground?** (AM.III.M.K3)

    A spring-loaded latch mechanism prevents the blades moving into the feather position when the engine is shut down on the ground. In the air, aerodynamic forces keep the propeller rotating fast enough that centrifugal force holds the blades unlatched so they can move to the feather position when oil pressure is taken out of the propeller.

15. **What is done to cause a hydromatic propeller to feather?** (AM.III.M.K3)

    High-pressure engine oil is directed into the propeller through the governor.

16. **What does the pilot do to change the RPM of an engine equipped with a constant-speed propeller when it is operating within the constant-speed range?** (AM.III.M.K4)

    The pilot moves the propeller pitch control. This changes the compression on the speeder spring inside the governor which moves the pilot valve. The pilot valve directs oil into or out of the propeller to change the pitch of the blades. The change in pitch changes the air load on the propeller which changes the RPM.

17. **What is the difference between a controllable propeller and a constant-speed propeller?** (AM.III.M.K4)

    Basically, it is the control system. A controllable-pitch propeller uses a manually operated oil valve to control the pitch, and a constant-speed propeller uses a governor to control the valve.

18. **When making a magneto check on an engine equipped with a constant-speed propeller, should the propeller control be in the low-pitch or the high-pitch position?** (AM.III.M.K4)
It should be in the low-pitch, high RPM position.

19. **What is meant by the beta range of operation of a turboprop propeller?** (AM.III.M.K5)
This is the mode of ground operation, and it includes starting, taxiing, and ground reverse operation.

20. **What controls the blade angle of a propeller when operating in the beta range?** (AM.III.M.K5)
When operating in the beta range, blade angle is controlled by the position of the power lever.

21. **Where do you find a list of the lubricants that are approved for use in a constant-speed propeller?** (AM.III.M.K6)
In the maintenance manual for the propeller.

22. **What is the extent of the repairs a mechanic with a Powerplant rating can make to a propeller?** (AM.III.M.K6)
Only minor repairs or minor alterations.

23. **Where can examples of acceptable repairs to aluminum alloy propeller blades be found?** (AM.III.M.K6)
In AC 43.13-1B, Chapter 8, Section 4, and in the propeller manufacturer's service manuals.

24. **Is it permissible to cold straighten a damaged aluminum alloy propeller blade to facilitate shipping it to a repair station?** (AM.III.M.K6)
No, this could cause hidden damage that may render the blade nonrepairable.

25. **What would be the classification of maintenance for shortening a propeller blade?** (AM.III.M.K6)
Major repair.

26. **May transverse cracks in a metal propeller blade be repaired?** (AM.III.M.K6)
No, a transverse crack of any size is reason for rejecting the blade.

27. **What damage to an aluminum alloy propeller blade can be repaired by a mechanic holding a Powerplant rating?** (AM.III.M.K6)
Small roughness, nicks, and scratches in the leading edge of the blades.

28. **How are small nicks removed from the leading edge of a propeller blade?** (AM.III.M.K6)
File them out with a fine file or stone, leaving a smooth contour.

29. How can you determine that a surface scratch in an aluminum alloy propeller blade is not actually a crack? (AM.III.M.K6)

Clean the damage out to a saucer-shaped depression and spray the area with a dye-penetrant liquid. Allow it time to seep into a crack if one is present, then wipe all of the liquid off the surface. Spray the area with a developer. If the damage is actually a crack, the developer will pull the penetrant out and it will form a visible line.

30. Who, or what facility, is authorized to perform major repairs to a propeller? (AM.III.M.K6)

An FAA-approved propeller repair station that is authorized for the specific propeller.

31. What checks and maintenance is a mechanic with a Powerplant rating allowed to make on a propeller? (AM.III.M.K6)

- Check the track of the propeller blades.
- Remove small nicks and scratches from the blades.
- Check the dynamic balance of a propeller.
- Lubricate the propeller.

32. What adjustment can a mechanic with a Powerplant rating make to a propeller governor? (AM.III.M.K6)

Adjust the maximum RPM stop.

33. What instrument is used to measure the blade angle of a propeller? (AM.III.M.K6)

A universal propeller protractor.

34. Where is the propeller protractor placed to measure the propeller blade angle? (AM.III.M.K6)

At a distance from the center of the propeller hub. This distance is specified in the propeller maintenance manual in terms of propeller blade stations.

35. What are the basic steps for removing a typical propeller? (AM.III.M.K7)

1. Remove the spinner and remove safety wire from the mounting bolts.
2. Support the propeller with a sling and mark the alignment between the propeller hub and the engine flange with a felt-tipped pen.
3. Unscrew the mounting nuts and protect the threads of the mounting studs or bolts as the propeller is removed.
4. Place the propeller on a cart for transport.

36. What alignment method is used to align a propeller to the engine in a position that reduces operating vibrations? (AM.III.M.K7)

Special studs or dowel pins align the propeller to the engine in the proper orientation.

37. Who is allowed to reduce the diameter of a type certificated propeller? (AM.III.M.K8)

An FAA-certificated propeller repair station with approval for the particular propeller.

38. Which type certificate data sheet (aircraft, engine, or propeller TCDS) specifies that a propeller can be installed on the aircraft in question? (AM.III.M.K8)

The aircraft TCDS specifies which propellers are approved for installation.

39. **What is meant by a slave engine with regard to propeller synchronization?** (AM.III.M.K9)
This is the engine in a multi-engine airplane whose RPMs follow those set on the master engine.

40. **How are propellers on a twin-engine airplane synchronized?** (AM.III.M.K9)
The propeller governor of the master engine is set to the desired RPM. A signal from the master engine governor is sent to the control box which sends a signal to a stepping motor. This motor adjusts the propeller governor and fuel control of the slave engine causing it to maintain exactly the same RPM as the master engine.

41. **How is ice prevented from forming on propeller blades?** (AM.III.M.K10)
A chemical anti-icing system is used. A mixture of isopropyl alcohol and ethylene glycol is pumped into a slinger ring on the back of the propeller hub. From there centrifugal force slings it out along the blades. Ice cannot form on the resulting slick surface.

42. **How is ice removed from propeller blades?** (AM.III.M.K10)
An electrothermal deicing system is used. Electrical heating elements are embedded in rubber boots that are bonded to the leading edges of the blades. Low-voltage DC flows from the propeller deicer control through brushes and slip rings to the heating elements. The timer sends current to one propeller for about 90 seconds and then to the other for the next 90 seconds. This allows the ice to form and then break loose and blow away.

43. **How can you determine that the electrically heated deicer boots on the propeller blades are working as they should?** (AM.III.M.K10)
Observe the loadmeter or ammeter to see if the proper current is flowing, and follow the sequence of boot heating by feeling with your hands to see which one is heating. The boots should all have a similar heat rise in the same length of time.

Risk Management

1. **What dangers are present when operating a propeller on the ground?** (AM.III.M.R1)
Propellers become almost invisible when spinning rapidly and may not be visible to individuals close to the airplane. Measures must be in place to prevent people from approaching the airplane when the engine is running. Propeller blast is also a potential danger that can cause damage to equipment or other aircraft that are behind the running engine. Before starting an engine, the aviation mechanic must use caution and understand what is behind the operating engine. In addition, the area below and in front of the propeller must be checked for gravel, small rocks, and other items that could become FOD and damage the propeller when the engine is started.

2. **What is the critical nature of propeller maintenance and inspection?** (AM.III.M.R2)
The loads on a spinning propeller are extremely high. Small damage and improper maintenance can lead to catastrophic damage to the aircraft due to failure of the propeller. Maintenance manual procedures must be followed at all times.

Skills

1. **Remove and replace a fixed-pitch propeller on an engine with a splined or tapered shaft. Explain the reason propellers are indexed to the crankshaft. Explain the reason for the snap ring in the propeller hub and demonstrate the application of the proper torque to tighten the retaining nut. Explain what to do if the front cone bottoms when the propeller is reinstalled.** (AM.III.M.S1)

2. **Remove and replace a constant-speed propeller on an engine specified by the examiner.** (AM.III.M.S1)

3. **Remove, inspect, and/or install a propeller governor.** (AM.III.M.S1)

4. **Check blade static tracking.** (AM.III.M.S2)

5. **Inspect a wooden propeller metal tipping.** (AM.III.M.S3)

6. **Perform a check of tip tracking a metal fixed-pitch propeller. Explain to the examiner the correct way to correct a minor out-of-track condition.** (AM.III.M.S3)

7. **Perform a dye-penetrant inspection to an area on an aluminum alloy propeller blade specified by the examiner.** (AM.III.M.S3)

8. **Inspect a propeller for damage and perform a repair to remove small surface damage.** (AM.III.M.S3)

9. **Remove a spinner from a constant-speed propeller and check the spinner bulkhead for cracks. Explain the proper type of repair to a cracked bulkhead. Explain the precautions that must be taken when reinstalling the spinner.** (AM.III.M.S3)

10. **Clean an aluminum alloy propeller.** (AM.III.M.S3)

11. **Demonstrate the use of a universal propeller protractor to measure the blade angle of a propeller specified by the examiner. Explain the reason for choosing the position to place the protractor.** (AM.III.M.S4)

12. **Perform a minor repair to a metal propeller blade.** (AM.III.M.S5)

13. **Select the proper lubricant for a propeller as specified by the propeller manufacturer.** (AM.III.M.S6)

14. **Demonstrate to the examiner the correct way to lubricate a propeller. Use the proper reference materials to find the lubricants that are to be used.** (AM.III.M.S6)

15. **Locate the procedures for balancing a fixed-pitch propeller.** (AM.III.M.S7)

16. **Adjust a propeller governor.** (AM.III.M.S8)

17. **Using the correct reference material, find out if there is any RPM range that is restricted for an engine-propeller combination specified by the examiner.** (AM.III.M.S9)

18. **Identify propeller range of operation on the assigned aircraft and propeller combination.** (AM.III.M.S9)

19. **Perform a 100-hour inspection on a propeller.** (AM.III.M.S10)

20. **Troubleshoot a turboprop propeller system.** (AM.III.M.S10)

21. **Determine what minor propeller alterations are acceptable using the propeller specifications, TCDS, and listings.** (AM.III.M.S11)

22. **Repair an anti-icing or de-icing system on a propeller.** (AM.III.M.S12)

23. **Demonstrate to the examiner the way to check and replace the brushes for an electrothermal propeller deicer.** (AM.III.M.S12)

24. **Demonstrate the correct way to check for proper operation of a propeller electrothermal deicer system.** (AM.III.M.S12)